Altering the Earth's Chemistry: Assessing the Risks

Sandra Postel

Worldwatch Paper 71
July 1986

Sections of this paper may be reproduced in magazines and news-papers with acknowledgment to Worldwatch Institute. The views expressed are those of the author and do not necessarily represent those of Worldwatch Institute and its directors, officers, or staff.

Table of Contents

Introduction

Throughout geologic time, plants and animals have altered and shaped conditions for life on earth. Some 2 billion years ago, green plants began adding oxygen to the atmosphere, a product of their use of sunlight to transform raw materials into food. This oxygen was deadly to the many organisms of the time that had evolved without it, which prompted British scientist James E. Lovelock to call its release the "worst atmospheric pollution incident that this planet has ever known." Fortunately, the oxygen buildup occurred slowly, and though many species undoubtedly were extinguished, the planet's ecosystems gradually accommodated the changed chemistry of the environment.[1]

Over the last two centuries—a mere instant of geologic time—human activities have altered the earth's chemistry in ways that may cause staggering ecological and economic consequences within our lifetimes or those of our children. Three stand out as particularly threatening and costly to society: risks to food security, to forests, and to human health. These risks arise from everyday human activities that collectively have reached a scale and pace capable of disrupting natural systems that evolved over millions of years.

Scientists now expect that a buildup in the atmosphere of certain carbon, nitrogen, and chlorine compounds will change the earth's climate more over the next 50 to 75 years than it has changed over the last 15,000 years. The most worrisome compounds are released from the combustion of fossil fuels (coal, oil, and natural gas), the clearing

I sincerely thank Gretchen Daily for her research assistance, and Dieter Deumling, Andrea Fella, Maureen Hinkle, William Kellogg, Justin Lancaster, Gene Likens, Alan Miller, Ralph Rotty, and Arthur Vander for their helpful comments on drafts of the manuscript.

and burning of forests, some intensive agricultural practices, and a few consumer products. In the atmosphere, these chemicals trap heat radiated from the earth that would otherwise escape more rapidly into space. The anticipated result is a global warming that will shift temperature and rainfall patterns worldwide, making crops in some key food-producing regions more vulnerable to heat waves, drought, and the loss of water supplies for irrigation. Maintaining food security under such an altered climate could require as much as $200 billion to adjust irrigation patterns alone. According to some estimates, though highly uncertain, the annual cost of climate change could approach 3 percent of the world's gross economic output, perhaps cancelling the benefits of economic growth.[2]

In many industrial countries, pollutants from fossil-fueled power plants, metal smelters, and automobiles are now stressing forests beyond their ability to cope. Trees covering at least 10 million hectares in Europe—an area the size of East Germany, or the U.S. state of Ohio—now show signs of injury linked to air pollutants or acid rain. While most exhibit the early stages of yellowing or foliage loss, millions are dead or dying. Researchers in West Germany project from current trends that forest damages in their country over the coming decades will average $2.4 billion per year. These damages include not only losses to forest industries, but of water quality and recreational values that healthy forests help maintain.[3] In some areas, including much of the eastern United States, fossil fuel pollutants may be diminishing the growth and productivity of trees while causing no visible damage at all.

People suffer from the same pollutants injuring forests. Despite impressive air quality gains made in industrial countries over the last two decades, many people are still exposed to harmful pollution levels. Fossil fuel pollutants may cause as many as 50,000 premature deaths in the United States each year.[4] Metals, which occur naturally in soils and also are released to the air during combustion, smelting, and other industrial processes, are of growing concern. Some are toxic to humans when ingested or inhaled in large enough amounts. Moreover, recent findings suggest that acid rain could magnify these

> **"By the time researchers document a marked change in climate, it will be irrevocable, and the consequences unavoidable."**

risks by releasing some potentially harmful metals that would otherwise remain bound up in soils.

Risks to health have also emerged with the thousands of synthetic chemicals created in just the last few decades. Exposed to these compounds through air, water, and food, the human body has had insufficient time to evolve special mechanisms to cope with them. Though many are linked to serious health effects, the total risks synthetic chemicals pose are unknown: Most have scarcely been tested for toxicity.

7

Much scientific uncertainty surrounds each of these threats, and more research is urgently needed. Yet waiting for a definitive picture of how each threat will unfold invites costly consequences and potential disasters. By the time researchers document a marked change in climate, it will be irrevocable, and the consequences unavoidable. Scientists know of no practical way to reverse the damage to forests and soils now spreading in Europe. Birth defects or cancer may only appear decades after exposure to the offending chemicals, and by then the health damage cannot be undone. Such irreversibility requires citizens and political leaders to act before the consequences of chemical pollution fully emerge.

An unsettling element of surprise also pervades environmental threats. Natural systems—including climate, forests, and the human body—may absorb stresses for long stretches of time without much outward sign of damage. A point comes, however, when suddenly conditions worsen rapidly. Scientists may anticipate such sudden changes—variously called "jump events," "thresholds," or "inflection points"—but rarely can they pinpoint when they will occur. As the scale and pace of human activities intensify, the risk of overstepping such thresholds increases.

Time figures prominently in assessing the risks of human-induced changes in the earth's chemistry. Given sufficient time to adjust, the earth and its organisms exhibit remarkable resilience in the face of change. Yet during that period of adjustment—whether to higher global temperatures, a blanket of air pollutants, acidic rain, or toxic chemicals—much suffering, economic loss, and social disruption may occur. As a dominant agent of change, humanity must confront the adverse consequences of that change, and protect this and future generations from them.

Agents of Change

In scarcely two centuries, industrial societies have altered basic chemical cycles that evolved gradually over many millennia. Just six elements—carbon, oxygen, nitrogen, hydrogen, phosphorus, and sulfur—comprise 95 percent of the mass of all living matter on earth. Since the supply of these elements is fixed, life depends on their efficient cycling through the atmosphere and the rocks, soils, waters, and organisms of the biosphere, a process called biogeochemical cycling. In recent years, researchers have learned that human activities have significantly disrupted these cycles, notably those of carbon, nitrogen, and sulfur.[5]

Since 1860, the combustion of fossil fuels has released some 185 billion tons of carbon to the atmosphere. Annual emissions rose from an estimated 93 million tons in 1860 to about 5 billion tons at present, a 53-fold increase. The bulk of these emissions occurred since 1950 as carbon releases from the rapid rise in oil-use added substantially to those from coal. (See Figure 1.) Earlier in this century, the clearing and burning of forests to make way for cropland and pasture contributed even more carbon to the air each year than fossil fuels did. Scientists estimate that between 1860 and 1980 forest clearing released to the atmosphere more than 100 billion tons of carbon. Today, land conversion—principally deforestation in the tropics—is estimated to cause a net release of between 0.6 billion tons and 2.6 billion tons of carbon annually, or between 12 and 50 percent of that released each year from fossil fuel combustion.[6]

Scientists voiced concern about this addition of carbon to the atmosphere as long as a century ago. Until fairly recently, many assumed that the oceans—the biggest reservoir in the carbon cycle—would

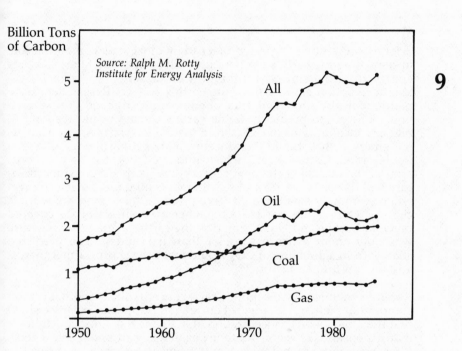

Billion Tons of Carbon

Source: Ralph M. Rotty
Institute for Energy Analysis

All

Oil

Coal

Gas

5

4

3

2

1

1950   1960   1970   1980

Figure 1: World Carbon Emissions from Fossil Fuels, 1950-84

remove this carbon added by human activities. Then in the late fifties, researchers atop Hawaii's Mauna Loa began to measure the concentration of atmospheric carbon dioxide (CO_2), and found that it was rising. Between 1959 and 1985, the annual average CO_2 concentration increased from 316 parts per million (ppm) to about 345 ppm, or 9 percent. With CO_2 levels prior to 1860 estimated at 260-270 ppm, human activity has increased the concentration of carbon dioxide by about 30 percent in just 125 years. About half of the CO_2 emitted to the atmosphere has been absorbed by the oceans and the biosphere. The release of carbon from combustion and deforestation thus appears to outpace the natural carbon cycle's ability to keep CO_2 levels constant.[7]

10 Scientists have long suspected that carbon dioxide plays a central role in regulating the earth's temperature. Like a one-way filter, CO_2 lets energy from the sun pass through it but absorbs the longer wavelength radiation emitted from the earth's surface. Researchers have mathematically modelled this phenomenon, dubbed "the greenhouse effect," to predict how the earth's climate would respond to higher concentrations of CO_2. From the models' results, a consensus has emerged that should the concentration of atmospheric CO_2 reach double preindustrial levels (which under existing trends will occur around the middle of the next century) the earth's average temperature will rise between 1.5 and 4.5 degrees centigrade. Such a change, while seemingly small, would have profound effects on the world's climate. During the last Ice Age, when vast sheets of ice covered much of Europe and North America, the earth's average temperature was only about 5 degrees colder than it is today. The predicted change from a doubling of CO_2 would make the earth warmer than at any time in human history.[8]

Recently scientists have begun to raise concerns about another carbon compound: methane. Studies of ancient air trapped in polar ice show that the atmospheric concentration of methane remained constant for many thousands of years, but then began rising around the year 1600. It has since more than doubled, and is now increasing at a rate of 1 to 2 percent per year. The exact cause of this increase remains uncertain. Most gaseous methane is produced by bacteria that decompose organic matter in oxygen-deficient environments. Bacteria in the digestive tracts of cows and the soils of rice paddies worldwide produce on the order of 140 million tons of methane annually, perhaps double the amount released from natural swamps and wetlands. Like carbon dioxide, methane acts as a greenhouse gas, trapping heat from the earth. Researchers estimate that methane's buildup in the atmosphere by the year 2030 could increase the global warming expected then from carbon dioxide by between 20 and 40 percent.[9]

Only a partial picture exists of the global nitrogen budget, but human activities clearly have altered the cycling of this key element as well. The combustion of fossil fuels releases oxides of nitrogen (NO_x) to the atmosphere through the oxidation of nitrogen in fuel and in the air of

"The predicted change from a
doubling of CO_2 would make the
earth warmer than at any time in
human history."

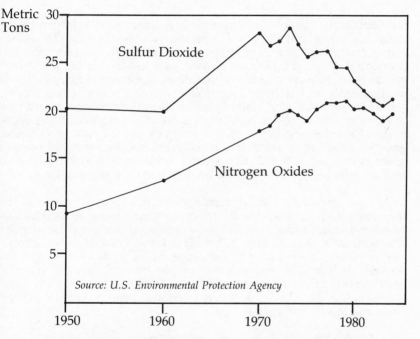

**Figure 2: U.S. Emissions of Sulfur Dioxide and
Nitrogen Oxides, 1950-84**

combustion chambers. Power plants, automobiles, and industries are
large emitters of these compounds. In the United States, NO_x emissions rose from 9.3 million tons in 1950 to 20.2 million tons in 1973,
and have since remained at roughly this high level. (See Figure 2.)
Current estimates place worldwide NO_x emissions from human activities equal to those from lightning, soils, and other natural sources
combined.[10]

Parts of the nitrogen cycle have also accelerated with the increasing
intensity of crop and livestock production. To meet growing world

food demands, farmers have applied large amounts of nitrogen-based fertilizers to the land and have stepped up meat production by raising cattle in feedlots. Scientists believe that the fertilization of soils, the concentration of animal wastes, and to a lesser degree, fossil fuel combustion, release substantial quantities of nitrous oxide, known to many people as laughing gas. Unlike the NO_x compounds, which settle or rain out after a relatively short life in the atmosphere, nitrous oxide may remain for a century or more. Concern about this compound has intensified with the discovery that it, too, is a greenhouse gas. Scientists estimate that the projected concentration of nitrous oxide in the year 2030 will increase the global warming by between 10 and 20 percent of that expected by then from carbon dioxide.[11]

Unlike nitrogen and carbon, sulfur maintains no major reservoir in the atmosphere, yet a portion of it cycles through the air as it moves between the land and sea. Each year volcanoes, sea spray, wetlands, and tidal flats release 90 to 125 million tons of sulfur to the atmosphere.[12] The greatest human influence on the cycle comes from industrial activity—mainly the combustion of coal and oil and the smelting of sulfur-bearing metallic ores. These sources emit the compound sulfur dioxide (SO_2). (Every two tons of sulfur dioxide emitted adds one ton of sulfur to the air.)

Worldwide, humanity's annual contribution of sulfur to the atmosphere now roughly equals that of all natural sources combined— about 100 million tons, essentially doubling the annual cycling of sulfur through the biosphere. In the United States alone, SO_2 emissions increased by 40 percent between 1950 and 1973, peaking at nearly 30 million tons. (See Figure 2.) As a result of oil price increases and the enactment of air pollution standards, SO_2 emissions declined through the late seventies and early eighties, though data for 1984 show a reversal of that trend. Nevertheless, emissions remain high: In eastern North America, sulfur output from human activities exceeds that from natural sources by a factor of at least ten.[13]

Once aloft, NO_x and SO_2 react with other chemicals in the lower atmosphere. NO_x and hydrocarbons, for example, are both emitted by the combustion of oil in automobiles. Under intense sunshine, they help form ozone, a principal ingredient in the "photochemical smog" that blankets many urban areas. These pollutants can migrate from cities to the countryside, sometimes carried great distances by prevailing winds. In many rural areas of Europe and North America, summer ozone concentrations now measure two to three times higher than natural background levels.[14]

13

Similarly, acid deposition forms from a complex set of reactions involving sulfur and nitrogen compounds. In the mid-1800s, English chemist Robert Angus Smith studied the precipitation falling around Manchester, England, and found higher sulfuric acid levels in town than in the surrounding countryside. Over the last several decades, however, acid rain has spread widely throughout rural areas in industrial countries. The increase in emissions of acid-forming pollutants, along with tall smokestacks designed to disperse them away from cities, converted acid rain from a localized urban problem to one regional and continental in scale.[15]

Monitoring of precipitation chemistry in Europe began in the 1950s and shows that the area receiving highly acidic rain and snow has spread from a central region of Belgium and the Netherlands to include nearly all of northwestern Europe. Long-term data are lacking for Eastern European countries, but the heavy burning of sulfur-rich brown coals likely has increased precipitation acidity there to levels higher than elsewhere in Europe. In North America, throughout a broad region extending across eastern Canada, south to Tennessee and the Carolinas, and west to Iowa and Missouri, the acidity of rainfall now averages at least ten times greater than would be expected in an atmosphere free of industrial pollutants.[16]

Industrial activities have also turned metals into troublesome pollutants. Metals occur naturally in soils and rock, and in the forms and concentrations found in nature, pose little hazard. However, with the growth of fossil fuel combustion, smelting, incineration, and other high-temperature processes, metal concentrations in the environment have increased markedly. For nearly a dozen metals, releases to the atmosphere from human activities now greatly exceed those from soils, volcanoes, and other natural sources. (See Table 1.) Emissions

Table 1: Estimated Annual Global Emissions of Selected Metals to the Atmosphere, circa 1980

Metal	Human Activity	Natural Activity	Ratio of Human to Natural Activity
	(thousand metric tons)		
Lead	2,000	6	333
Zinc	840	36	23
Copper	260	19	14
Vanadium	210	65	3
Nickel	98	28	4
Chromium	94	58	2
Arsenic	78	21	4
Antimony	38	1	38
Selenium	14	3	5
Cadmium	6	0.3	20

Source: James N. Galloway et al., "Trace Metals in Atmospheric Deposition: A Review and Assessment," *Atmospheric Environment*, Vol. 16, No. 7, 1982.

of cadmium have been increased 20-fold, and of zinc 23-fold. The use of lead in gasoline, which began in the twenties, has boosted lead emissions worldwide to 2 million metric tons annually—333 times greater than estimated releases from natural sources.[17]

Like emissions of acid-forming pollutants, metals return to earth and are deposited in soils, streams, and lakes. While lead concentrations in uncontaminated soils typically range between 10-50 ppm, scientists have measured levels in urban areas of the northeastern United States ranging between 100-800 ppm. Similarly in Taiwan, which has one of the highest traffic densities in the world, lead concentrations in some areas exceed natural levels by 25 times. Scientists do not know the regional or global extent of metal deposition, since it has not been monitored extensively. But after studying the accumulation of metals in forests of the Solling mountains in central Germany, two German scientists concluded that "since the Solling is a typical rural mountain area, far from large industrial settlements and major traffic lines...there is probably no land surface in central Europe where the local balance for most heavy metals on the ecosystem level is not strongly influenced, or dominated, by atmospheric pollution."[18]

Moreover, a team of North American researchers found from a literature survey covering nearly 300 sites worldwide that metal deposition rates in rural areas were 10 to 100 times greater than in the remote North Atlantic. In urban areas, deposition rates were between 100 and 10,000 times greater. They concluded that mercury and lead "are now being deposited in some areas at levels toxic to humans" and cadmium, copper, mercury, lead, and zinc "at levels toxic to other organisms."[19]

Besides altering the cycling of natural elements such as carbon, nitrogen, sulfur, and metals, society has introduced to the environment over the last half-century thousands of substances that have no natural counterparts. Their early creators probably never imagined that the use of these chemicals might severely damage natural systems. Yet in the early seventies, scientists warned that one family of synthetic compounds—the chlorofluorocarbons—could destroy the life-protecting layer of ozone in the upper atmosphere.

Ironically, ozone—the same chemical that in the lower atmosphere forms irritating urban smog—in the upper atmosphere performs a vital function. It absorbs ultraviolet (UV) radiation from the sun, which if allowed to reach the earth would cause skin cancers, crop damage, and other harmful effects. Once aloft, the chlorofluorocarbons (CFCs) migrate to the upper atmosphere, where the sun's intense rays break them down, releasing atoms of chlorine. This chlorine in turn drives a series of reactions that destroy ozone. Largely as a result of worldwide CFC emissions, stratospheric concentrations of chlorine are now more than twice natural levels.[20]

Production of CFC-11 and CFC-12, the most worrisome members of the CFC family, rose steadily from the early thirties to the early seventies as demand grew for their use as propellants in aerosol cans, as foam-blowing agents, and as coolants for refrigerators. Production declined from the mid-seventies through 1982 after a number of industrial nations responded to scientists' warnings by banning or restricting aerosol uses of CFCs. Yet because of increasing demands for CFC products in Third World countries and for unrestricted CFC uses in industrial countries, this downward trend appears to have reversed. Estimated production (excluding the Soviet Union, Eastern Europe, and China) rose 16 percent between 1982 and 1984.[21]

Projections of the potential effect of CFC releases have varied widely over the last six years. Assuming modest emissions growth-rates for related atmospheric gases and a 3 percent per year increase in CFC emissions, the U.S. National Aeronautics and Space Administration has projected a 10 percent depletion of the ozone layer by the middle of the next century. According to a study by the U.S. Environmental Protection Agency (EPA), such a depletion could result in nearly 2 million additional skin cancer cases each year, damage to materials such as plastics and paints worth as much as $2 billion annually, as well as incalculable damage to crops and aquatic life. Studies also show that exposure to UV rays may suppress the human immune system, potentially making people more vulnerable to disease. The higher doses of UV radiation resulting from ozone depletion would likely magnify this damaging effect. Moreover, concern about the pace and predictability of ozone depletion has heightened recently with findings of roughly a 40 percent decrease in the ozone layer above Antarctica during October, shortly after sunlight reappears following the continent's cold, dark winter. Scientists had not anticipated such a loss, and whether it portends a more-rapid-than-expected depletion of the ozone layer globally remains unknown.[22]

Chlorofluorocarbons also add to the threat of climate change, both indirectly by their attack on the ozone layer, and directly, because they act as greenhouse gases. The expected ozone depletion will alter the energy budgets of the upper and lower atmospheres, tending to warm the earth. Since CFCs themselves effectively trap heat radiated from the earth, their buildup in the atmosphere will add substantially to the greenhouse warming. In October 1985, scientists from 29 nations meeting in Villach, Austria, concluded that the climate-

"A 10 percent depletion of the ozone
layer could result in nearly 2 million
additional skin cancer cases each
year."

altering potential of greenhouse gases other than CO_2—which include methane, nitrous oxide, and the CFCs—"is already about as
important as that of CO_2." Taken together, the rising concentrations
of carbon dioxide and all other greenhouse gases could lead to the
equivalent of a doubling of CO_2 over preindustrial levels by "as early
as the 2030s."[23]

17

The long-term effects of the thousands of other synthetic chemicals
applied to croplands, emitted from factories, and dumped as waste
onto the land are largely unknown. In 1962, Rachel Carson's *Silent
Spring* alerted the world to the dangers posed by the proliferation of
these compounds. She wrote that they "come from our laboratories in
an endless stream; almost five hundred annually find their way into
actual use in the United States alone. The figure is staggering and its
implications are not easily grasped—500 new chemicals to which the
bodies of men and animals are required somehow to adapt each year,
chemicals totally outside the limits of biologic experience."[24]

As with many industrial products, the benefits of using chemicals are
easier to quantify than the costs. Pesticides, for example, have helped
control such dreaded diseases as malaria, bubonic plague, typhus,
and sleeping sickness, saving many millions of lives. They kill insects
that can devastate crops, and thus have arguably helped reduce hun-
ger and avert famine. Initially, concerns about their dangers were
limited to acute toxic effects of the sort that killed the unwanted pests.
Yet research has since revealed that pesticides and other chemicals
pose serious and often insidious long-term risks. Many degrade
slowly, lingering in soils or water for years after their release. Many
are volatile, and are transported great distances through the air. They
may increase in concentration as they work their way through the
food web, possibly reaching toxic concentrations in higher-level consumers. Most disturbing, many may harm both ecosystems and people exposed to minute concentrations of them for extended periods of
time.[25]

Findings on the long-term effects of pesticides took society by surprise. Few people could have known when the pesticide DDT came
into widespread use in the early forties that it would interfere with

18

the formation of normal eggshells in peregrine falcons, bald eagles, and other predatory birds, thus nudging these species toward rarity and extinction. Still fewer would have guessed that DDT would find its way to penguins in the Antarctic; that within three generations, most Americans—including unborn babies—would have measurable quantities of DDT in their blood and fat; or that more than a decade after its ban from use in the United States, DDT would still be found in carrots and spinach sold in San Francisco supermarkets.[26]

How extensively changes in the earth's chemistry will affect people and natural systems remains unknown. With built-in mechanisms of checks and balances, the biosphere tends toward a steady state, much as the human body maintains a constant internal temperature regardless of the temperature outdoors. Yet any self-regulating system can be so perturbed by external stresses that it destabilizes and loses its ability to function. Walter Orr Roberts of the National Center for Atmospheric Research expresses the well-grounded fear that humanity, "through the advancing power of science and technology, can overwhelm the curative and restorative forces at play...."[27]

Risks to Food Security

Over the centuries, farmers have geared their cropping systems to nature's normal offering of rain and warmth. Departures from these seasonal conditions can severely undermine crop production, farmers' livelihoods and, ultimately, food security. If current trends continue, agriculture's battle with the weather will intensify over the coming decades. Because of the rising concentrations of carbon dioxide, chlorofluorocarbons, and other greenhouse gases in the atmosphere, scientists expect a marked change in the earth's climate over the next 50 to 75 years. Existing models cannot capture all of the complexities of the world's climate, nor can they predict precisely the changes in temperature and rainfall that will occur in specific regions. Yet they clearly indicate the need for some major and costly adjustments to maintain global food security.[28]

> "Any self-regulating system can be
> so perturbed by external stresses that it
> destabilizes and loses its ability to
> function."

Although the climate will change gradually as the concentrations of greenhouse gases increase, most modelers focus their predictions on what will occur from the equivalent of a doubling of carbon dioxide over preindustrial levels. They generally agree that temperatures will rise everywhere, though by greater amounts in the temperate and polar regions than in the tropics. Since a warmer atmosphere can hold more moisture, average precipitation worldwide is expected to increase by 7 to 11 percent. In many regions, however, this additional rainfall would be offset by higher rates of evaporation, causing soil moisture—the natural water supply for crops—to decrease.[29]

19

Recent model results indicate a substantial summertime drying out of the mid-continent, mid-latitude regions of the Northern Hemisphere. Soil moisture for summer crop production would diminish in large grain-producing areas of North America and the Soviet Union. Together Canada and the United States account for more than half of the world's cereal exports, and the United States alone accounts for 72 percent of the world's total exports of corn. (See Table 2.) In large portions of these areas, lack of water already limits crop production. A drier average growing season, along with more frequent and severe heat waves and droughts, could cause costly crop losses in these major breadbaskets. As a rule of thumb, for example, yields of corn in the United States drop 10 percent for each day the crop is under severe stress during its silking and tasseling stage. Thus, five days of temperature or moisture stress during this critical period, which would likely occur more frequently in much of the U.S. Cornbelt with the anticipated climate change, would reduce yields by half.[30]

While some key food-producing regions may dry out, prospects for expanding production in other areas could improve. Warmer and wetter conditions in India and much of Southeast Asia might increase rice production in these areas. The picture remains unclear for Africa. But reconstructions of the so-called Altithermal period some 4,500 to 8,000 years ago, when summertime temperatures were higher than at present, suggest that northern and eastern Africa could get substantially more rainfall. If so, average flows of the Niger, Senegal, Volta, and Blue Nile rivers would increase, possibly aiding the expan-

Table 2: Share of World Cereal Exports from Major Countries Where Summer Moisture Is Expected to Decrease, 1984

Country	Corn	Rice	Wheat	All Cereals
		(percent)		
Canada	1	0	19	11
Soviet Union	0	0	2	1
United States	72	17	37	44
Total	73	17	58	56

Source: United Nations Food and Agriculture Organization, *1984 FAO Trade Yearbook* (Rome: 1985); see text for sources on climate change.

sion of irrigation. In northern latitudes, higher temperatures and milder winters might open vast tracts of land to cultivation. Agricultural production in Canada, northern Europe, and the Soviet Union might expand northward.[31]

Unfortunately, shifting crop production to areas benefiting from climate change will not only be costly, but will meet with some serious constraints. Thin, nutrient-poor soils cover much of northern Minnesota, Wisconsin, and Michigan. A northward shift of the U.S. Cornbelt in response to higher temperatures would thus result in a substantial drop in yield. Poor soils will also inhibit successful northward agricultural migrations in Scandinavia and Canada. It would take centuries for more productive soils to form. Though during the Altithermal period the present desert regions of North Africa were savannas suited for grazing, these lands also would require a long time to regain their former fertility.[32]

Low-lying agricultural areas face the threat of a substantial rise in sea level from the altered climate. Since water expands when heated, oceans will rise with the increase in global temperature. Warmer temperatures will also melt mountain glaciers and parts of polar ice sheets, transfering water from the land to the sea. From the global warming expected by the middle of next century, sea levels could increase as much as one meter, threatening large tracts of agricultural lowlands—where much of the world's rice is grown—with inundation. Of particular concern are the heavily populated, fertile delta regions of the Ganges River in Bangladesh, the Indus in Pakistan, and the Yangtze in China.[33]

The productivity of major food crops will respond not only to changes in climate, but directly to the higher concentration of CO_2 in the atmosphere. Carbon dioxide is a basic ingredient for photosynthesis, the process by which green plants transform solar energy into the chemical energy of carbohydrates. Experiments suggest that as long as water, nutrients, and other factors are not limiting, every 1 percent rise in the CO_2 concentration may increase photosynthesis by 0.5 percent. Greenhouse operators have long taken advantage of this fertilizing effect by setting the CO_2 concentration in greenhouse air two to three times higher than that in today's natural atmosphere. Plants in a CO_2-enriched atmosphere also use water more efficiently since the leaf openings through which they transpire water narrow. Though few field studies have tested how major food crops would respond to higher CO_2 levels, researchers expect that, other things being equal, slight yield increases would occur.[34]

Other factors, however, could offset potential gains in yield. Crops might need more nitrogen and other nutrients to achieve the greater productivity made possible by higher CO_2 levels. Damage from insect pests could increase, since the warmer climate would likely enhance insect breeding. Yields of corn—the crop probably most vulnerable to the anticipated climate changes—could suffer from greater competition from weeds. Corn differs from wheat, rice, and most other major food crops in the way it carries out photosynthesis and, as a result, will not benefit as much from the fertilizing effect of higher CO_2 concentrations. Many weeds, however, will so benefit, and their greater ability to compete for water and nutrients could reduce corn production.[35]

Whatever the outcome for individual regions, adapting to climate change will exact heavy costs from governments and farmers. The expensive irrigation systems supplying water to the 270 million hectares of irrigated cropland worldwide were built with present climatic regimes in mind. These irrigated lands account for only 18 percent of total cropland, yet they yield a third of the global harvest. Irrigated agriculture thus plays a disproportionately large role in meeting the world's food needs. Shifts in rainfall patterns could make existing irrigation systems—including reservoirs, canals, pumps, and wells—unnecessary in some regions, insufficient in others. Moreover, seasonal reductions in water supplies because of climate change could seriously constrain irrigated agriculture, especially where competition for scarce water is already increasing.[36]

A look at one key food-producing region—the western United States—highlights how costly climate change could be. Though by no means conclusive, climate models suggest that much of the western United States could experience a reduction in rainfall along with the rise in temperature. Since rates of precipitation and evaporation largely determine any region's renewable water supply, supplies in the West would diminish. Assuming a 2 degree increase in temperature and a 10 percent decrease in precipitation, Roger Revelle and Paul Waggoner show that supplies in each of seven western river basins would be reduced by 40 to 76 percent. Such reductions would create severe imbalances in regional water budgets. (See Table 3.) Projected water consumption for the year 2000 would exceed supplies available under the current climate only in the Lower Colorado region. With the assumed climate change, however, consumption in the year 2000 would exceed the renewable supply in four regions, with local shortages likely occurring in the other three. The supply would fall 73 percent short of demand in the Rio Grande region, 64 percent in the Lower Colorado, 39 percent in the Upper Colorado, and 15 percent in the Missouri region.[37]

Since agriculture is by far the biggest consumer of water, balancing regional water budgets would likely require that irrigation cease on a substantial share of cropland. Such a trend is now under way in portions of the Lower Colorado region, where consumption already exceeds the renewable supply. Correcting the large imbalances resulting from such an altered climate could require that as many as 4.6

Table 3: Water Supplies under Present and Postulated Climate, Western United States

Water Resources Region	Average Annual Supply			Ratio of Demand in Year 2000 to Altered Supply
	Present Climate	Altered Climate[1]	Change	
	(billion cubic meters/year)		(percent)	
Missouri	85.0	30.7	−64	1.2
Arkansas-White-Red	93.5	43.2	−54	0.4
Texas Gulf	49.2	24.7	−50	0.7
Rio Grande	7.4	1.8	−76	3.7
Upper Colorado	16.4	9.9	−40	1.7
Lower Colorado	11.5	5.0	−57	2.7
California	101.8	57.1	−44	0.7
All 7 Regions	350.9[2]	165.3[2]	−53	0.9

[1]Assumes a 2 degree centigrade temperature increase and a 10 percent reduction in precipitation. [2]Does not equal sum of column because a portion of Lower Colorado flow is derived from Upper Colorado.

Source: Roger R. Revelle and Paul E. Waggoner, "Effects of a Carbon Dioxide-Induced Climatic Change on Water Supplies in the Western United States," in National Research Council, *Changing Climate* (Washington, D.C.: National Academy Press, 1983).

million hectares be taken out of irrigation in those seven western U.S. regions—roughly 35 percent of their existing irrigated area.[38]

A reduction of that magnitude would have high costs, measured either by the capital investments in dams, canals, and irrigation systems rendered obsolete, or by the replacement value of that irrigation infrastructure. Investment needs for expanding irrigation vary widely, but assuming a range of $1,500 to $5,000 per hectare, replacement costs could total $7 billion to $23 billion in the United States alone. Worldwide, maintaining food security under the altered climate would likely require new irrigation systems beyond those that would be added anyway as world food needs increased. If such additional systems were needed for an area equal to 15 percent of existing irrigated area, climate change could carry a global price tag of $200 billion for irrigation adjustments alone.[39]

The need for new drainage systems, flood control structures, cropping patterns, and crop varieties would greatly magnify the costs of adapting to a changed climate. According to some ballpark estimates, the annual cost of a greenhouse gas-induced warming of 2.5 degrees centigrade could amount to 3 percent of the world's gross economic output. Much of this cost would result from the loss of capital assets in agriculture. Poorer countries would have the most difficulty adapting, and since food production typically comprises a relatively large share of their incomes, their people would suffer disproportionately. Moreover, as climate expert William W. Kellogg points out, the need to adapt to climate change will arise "against a backdrop of increased world population, increased demands for energy, and depletion in many places of soil, forests, and other natural resources." The disruptions wrought by a changing climate may thus bring new pockets of famine, losses of income, and the need for huge capital investments that many countries will find difficult to finance.[40]

While no other human-induced alteration of the earth's chemistry poses risks to agriculture as great as climate change, at least one other exacts substantial costs as well. Scientists have documented that air pollution causes significant losses in crop output. By far the most destructive pollutant to crops is ozone, causing an estimated 90 percent of the crop damage linked to air pollution in the United States. Highly toxic to plants, and not always causing visible injury, ozone is quietly reducing crop yields in many agricultural regions.[41]

Ozone levels in many rural areas of the United States—including the Cornbelt, the Appalachian states, the Southeast, and the Southern Plains—now average between 1.5 and 2 times greater than natural background levels. Peak concentrations climb several times higher. Using county-level data on crop production and ozone concentrations, a 1984 study by the U.S. Office of Technology Assessment estimates that in 1978 ozone caused a $2 billion loss in U.S.

agricultural productivity. Yields of corn were reduced by 2.5 percent, wheat by 6 percent, soybeans by 13 percent, and peanuts—a crop highly sensitive to ozone—by 24 percent. An analysis by the EPA suggests that reducing current ozone levels by 40 percent would yield economic benefits to crop producers and consumers totaling $2.8 billion per year.[42]

25

Damage to crops from air pollutants is not as well documented elsewhere. Substantial losses undoubtedly occur in Europe, where much farmland lies within several hundred kilometers of major urban and industrial areas. To provide a rough estimate of potential damages, one study for the European Economic Community reasonably assumes that air pollutants are at levels harmful to plants in 40 percent of the Community's agricultural area, and that as a result, yields of winter wheat, barley, and potatoes are reduced by 10 percent. Based on the 1981 harvest, crop output valued at $1 billion would be lost annually. Even less data is available for Third World countries. With pollution control efforts there lagging behind those in industrial countries, however, damages from air pollutants are no doubt occurring, and are bound to worsen if emissions from the growing numbers of power plants, factories, and automobiles are not controlled.[43]

Loss of Forests

Growing threats to forests from changes in the chemistry of the atmosphere pose another set of potentially costly consequences during the coming decades. In the autumn of 1983, the West German Ministry of Food, Agriculture and Forestry galvanized both scientists and the citizenry with an unsettling finding: 34 percent of the nation's trees were yellowing, losing needles or leaves, or showing other signs of damage. Preliminary evidence pointed to air pollution and acid rain as contributing, if not the leading, causes. A more thorough survey in 1984 confirmed that the unusual tree disease was spreading. Foresters found that trees covering half of the nation's 7.37 million hectares of forests were damaged, including two-thirds of those in the southwestern state of Baden Württemberg, home of the fabled Black Forest.[44]

Spurred by West Germany's alarming discovery, other European nations took action to assess the health of their own forests. Different methods of surveying and estimating damage were used in various countries, so the results are not strictly comparable. Nonetheless, the assessments collectively show that trees covering 8 percent of Europe's 136 million hectares of forest exhibit signs of injury. (See Table 4.) The key symptoms for the conifer species—the hardest hit—parallel those found in West Germany: yellowing of needles, casting off of older needles, and damage to the fine roots through which trees take up nutrients. Forest injury pervades the band of central Europe bounded by latitudes 46 and 53 degrees North. In at least a half-dozen countries—Austria, Czechoslovakia, Luxembourg, the Netherlands, Switzerland, and West Germany—a quarter to half the forested area is damaged.[45]

National estimates in some cases belie the extent of damage in specific regions. Total damage in Sweden is placed at about 4 percent, but an estimated 20 percent of the forested area in the south is affected. In 1984, foresters in France surveyed portions of the French Jura and Alsace-Lorraine, adjacent to West Germany's Black Forest, and found that more than a third of the trees were injured, at least 10 percent of them severely. Indeed, the alpine region spanning portions of Austria, France, Italy, Switzerland, and West Germany exhibits the worst damage. Swiss officials have warned that avalanches and landslides resulting from the loss of tree cover will damage houses and farms, and may force people to evacuate some areas. Estimates for the Eastern European countries include severely damaged and dead trees, but probably not all those exhibiting early stages of needle loss and yellowing. Unfortunately, no forest damage data are available for Greece, Ireland, Portugal, and Spain.[46]

Unusual tree injury in North America appears much less extensive than in central Europe. In the high-elevation forests of the eastern mountain ranges, red spruce trees are undergoing a serious die-back—a progressive thinning from the outer tree crown inward. On a closely studied peak called Camels Hump in Vermont, researchers found that half the spruce trees have died. A 1982 survey indicated

Table 4: Estimated Forest Damage in Europe, 1985

Country	Total Forest Area	Estimated Area Damaged	Portion of Total Area Damaged
	(thousands of hectares)		(percent)
Austria	3,754	910	24
Belgium	616	111	18
Czechoslovakia	4,600	1,250	27
East Germany	2,900	350	12
France	15,075	—[1]	—[1]
Hungary	1,600	176	11
Italy	6,363	400	6
Luxembourg	82	42	51
Netherlands	309	138	45
Norway	8,330	400	5
Poland	8,677	600	7
Sweden	26,500	1,000	4
Switzerland	1,200	408	34
West Germany	7,371	3,824	52
Yugoslavia	9,500	1,000	11
Other	39,087	—	—
Total	135,964	10,609	8

[1]Surveys covered selected regions; see text.

Sources: *Allgemeine Forstzeitschrift*, No. 46, Munich, West Germany, November 16, 1985; West Germany figures from Federal Ministry of Food, Agriculture and Forestry, "1985 Forest Damage Survey," Bonn, West Germany, October, 1985; Italy figures from U.N. Food and Agriculture Organization, Forestry Department, "Long-Range Air Pollution: A Threat to European Forests," *Unasylva*, Vol. 37, 1985.

that red spruce are declining in a variety of forests throughout the Appalachian Mountains.[47]

More subtle signs of ill health come from the U.S. Forest Survey's discovery that pine trees in a broad region of the Southeast grew 20 to 30 percent less between 1972 and 1982 than they did between 1961 and 1972. These pines are important commercially, and sustained growth declines will reduce the amount of timber available for future harvests. In a November 1985 report, Forest Service analysts state that the net annual growth of softwood timber in the Southeast "has peaked and turned downward after a long upward trend." Though less well documented, unexpected growth declines appear to have occurred throughout the Appalachians, extending north into New England. In written testimony presented to the U.S. Senate in February 1984, soil scientist Arthur H. Johnson draws an unsettling parallel by noting that similar growth reductions preceded the "alarming incidences" of forest damage in Europe.[48]

Hundreds of scientists in the affected countries continue to search for the cause of this unprecedented forest decline. Collectively they offer a bewildering array of hypotheses, attesting to the difficulty of unraveling a mystery within a complex natural system. Most agree, however, that air pollutants—probably combined with natural factors, such as insects, cold, or drought—are a principal cause. Explanations focus on acid rain, gaseous sulfur dioxide, nitrogen compounds, heavy metals, and ozone, which singly or in combination cause damage variously through the foliage, forest soils, or both.[49]

A key symptom of tree injury—yellowing of needles—typically results from a deficiency of one or more nutrients. Among the most vital to a tree's health and productivity are calcium, magnesium, and potassium. Any one of these can become deficient in a tree either because the soil is lacking that particular element, the tree's roots cannot absorb it, or it is being leached from the tree's foliage faster than the tree can take it up from the soil.[50]

Both direct attacks to the foliage by air pollutants and acid rain, and indirect damage through changes in forest soils can alter the balance of nutrients and other critical metabolic functions within trees. In combination with natural stresses, both pathways probably con-

tribute to the tree disease and death spreading throughout central
Europe. Changes in the soil, however, pose the most troubling prospects for the future. A reduction in pollutant emissions would improve air quality and curb acid rain almost immediately, thus
allowing trees suffering from direct attacks on the foliage to recover
fairly quickly. Damage to soils from acidification, however, would be
irreversible for the near future, since soils would take decades or
centuries to recover.[51]

Evidence from a severely damaged forest in Eastern Europe suggests
that alterations in the soil are indeed taking place. Large portions of
the Erzgebirge mountains northwest of Prague, Czechoslovakia,
now resemble a wasteland. Near the industial city of Most, where
power plants burn high-sulfur coal, sulfur dioxide concentrations
average 112 micrograms per cubic meter, much higher than in most
industrial areas, and 13 times higher than in a seemingly undamaged
rural forest about 160 kilometers to the southeast. Peak concentrations register several times higher than the average. The nu-
merous dead and dying trees in this industrial region may thus be
succumbing to the "classic smoke injury" known to occur near large
sources of uncontrolled pollution.[52]

Detailed measurements of the chemistry of runoff from the Erzgebirge mountain forest also suggest, however, that acidification has
profoundly altered the soil's ability to support a forest. Czech geo-
chemist Tomas Paces found that losses of the nutrients magnesium
and calcium from the damaged forest averaged, respectively, 6.8 and
7.5 times greater than from the undamaged rural forest. (See Table 5.)
Less than half of these increased nutrient losses can be explained by
the higher rates of precipitation and thus of atmospheric chemical
inputs in the damaged forest. Runoff of aluminum, which normally
remains bound up in soil minerals, was 32 times greater than from the
undamaged forest. With the loss of calcium and other elements that
can buffer incoming acidity, aluminum mobilizes to serve as the
buffering agent. Scientists have learned that soluble forms of aluminum can be toxic to trees, damaging their roots and preventing
them from picking up vital nutrients from the soil. Finally, outputs of
nitrate from the damaged forest exceeded those from the undamaged

Table 5: Chemicals in Runoff from Forested Watersheds, Czechoslovakia, 1976-1982

Chemical	Undamaged Rural Forest[1]	Damaged Forest[2]	Ratio of Damaged to Undamaged
	(kilograms/hectare/year)		
Potassium	1.9	6.8	3.6
Magnesium	3.8	26.0	6.8
Calcium	9.9	74.0	7.5
Sulfate	9.0	96.0	10.7
Nitrate	0.6	12.0	20.0
Aluminum	0.1	3.2	32.0

[1]Average for 7 years, 1976-1982. [2]Average for 5 years, 1978-1982.

Source: Tomas Paces, "Sources of Acidification in Central Europe Estimated from Elemental Budgets in Small Basins," Nature, May 2, 1985.

forest by a factor of 20. Paces believes this reflects the damaged forest's inability to properly recycle nitrogen—a loss of basic ecosystem function.

Forests in the industrial regions of Eastern Europe have borne inordinately heavy pollutant loads over the last few decades. Few forests outside these regions have so drastically collapsed. Yet ecological theory firmly supports the possibility of more widespread destruction as chemical stress on forests accumulates over time. According to C.S. Holling of the University of British Columbia, natural systems may so successfully absorb stress that for long periods change occurs very slowly. Eventually, however, systems may reach a stress point, and "a jump event becomes increasingly likely and ultimately inevitable." Paces of Czechoslovakia believes such a threshold effect may occur with forests subject to soil acidification. Where and when this "inflection point" will be reached is not known, yet Paces says "its existence is suggested by the fact that the acidification of soils in

Central Europe has proceeded for decades whereas the die-back of forests is a relatively fast phenomenon which takes only a few years."[53]

Substantial economic losses already are occurring from the existing level of pollution stress on forests, and they will magnify greatly if the prospects of large-scale forest decline become reality. The Czechoslovakian Academy of Sciences estimates the cost of acid pollution at $1.5 billion annually, with forest damage accounting for much of the total. In West Germany, foresters are now harvesting dead and dying trees before their time, which both increases forest management costs and leads to timber surpluses that depress market prices. Professor H. Steinlin of Albert-Ludwigs University in Freiburg sees little potential for West German forest industries to stabilize the market by increasing exports, especially if forest damage simultaneously creates wood surpluses in Austria, Switzerland, and other neighboring countries. With domestic markets saturated and several countries trying to increase exports, lumber prices could drop sharply, causing severe economic losses to forest owners.[54]

Sometime in the future, when trees cut prematurely would otherwise have been harvested, a period of shortages and rising wood prices could occur. Projecting from current trends, one study by researchers at the Technical University of Berlin estimates that German forest industries will suffer monetary losses averaging $1 billion annually through the year 2060. Yet besides supplying timber, healthy forests help protect the quality of streams and groundwater supplies, control the erosion of soils, and provide recreational enjoyment for both Germany's citizens and tourists. Adding in these projected losses, the Berlin researchers estimate that the total cost of forest damage over the next several decades will average $2.4 billion per year.[55]

In the United States, field and laboratory experiments, combined with the findings of greatly reduced tree growth, strongly suggest that ozone—already known to be diminishing crop yields—is reducing the productivity of some commercial forest species. Researchers at Cornell University subjected four species—white pine, hybrid poplar, sugar maple, and red oak—to a range of ozone concentrations spanning those typically found in the United States. In all four species, net photosynthesis, which is a measure of a tree's growth, decreased linearly with increasing ozone concentrations. This held true even at ozone levels now typical of rural areas. In no case were the reductions in photosynthesis accompanied by visible damage to the foliage. Thus, even with no outward sign of injury, trees covering large regions are very likely losing vigor and growing slower. As researchers Peter B. Reich and Robert G. Amundson point out, growth reductions of just 1 to 2 percent per year amount to a sub-stantial loss of timber over a tree's lifetime.[56]

In the Third World, uncontrolled clearing of forests is presently a much greater cause for concern than forest damage from air pollution or acid rain. Yet the health and productivity of forests in developing countries are bound to diminish in the future if pollution from power plants, industries, and motor vehicles continues to increase. Damage may first appear in the vicinity of large polluted urban centers, as it did, for example, surrounding West Germany's Ruhr Valley during that nation's earlier stages of industrialization. Indeed, trees re-portedly are dying along heavily travelled corridors in Mexico City. As in industrial countries, however, damage may spread to rural areas as pollution emissions increase, as trees remain exposed to pollutants for longer periods of time, or as soils acidify.[57]

Chronic pollution stress—whether from ozone, acid rain, sulfur di-oxide, nitrogen compounds, or metals—now places a substantial share of the industrial world's forests at risk. In just one year, forest damage in West Germany jumped from 34 percent to 50 percent. The 1985 damage survey showed just a slight increase, to 52 percent, perhaps because of weather conditions favorable for the forests.[58] No one knows how much the forest damage in all of Europe—now at least 8 percent—will increase. Nor does anyone know how many of the injured trees will eventually die, or when thresholds may be reached beyond which forest damage rapidly worsens. Whether the unexplained growth reductions in eastern U.S. forests portend a similar decline there also remains unknown. Meanwhile, with each passing year of continued pollution stress, the costs of lost forest productivity mount, as do the risks of more extreme forest decline and death.

Threats to Human Health

In 1775, during England's "industrial revolution," epidemiologist Percival Pott identified the first environmental carcinogen. He found surprisingly high rates of scrotal cancer among British chimney sweeps, and uncovered the cause to be their unusually high exposure to soot, a byproduct of combustion.[59] Since then, the health hazards of environmental pollutants have spread widely to the general population. The same fossil fuel pollutants damaging forests and crops also harm people. Metals released into the environment have become a growing cause for concern. Most recently, the proliferation of synthetic chemicals applied to crops, dispersed into the air, and disposed of on land, has added new dimensions to environmental health risks. Sacrifices of human health comprise part of the price of altering the earth's chemistry.

With the growth of cities and industries over the last century, pollution from the burning of fossil fuels amassed in ever greater quantities. Many cities became cloaked in a persistent pall of smoke and haze. Intense pollution episodes in the Meuse Valley of Belgium in 1930, in Donora, Pennsylvania, in 1948, and in London in 1952 drove home the acute health hazards of severely polluted air. During these periods of high sulfur dioxide and particulate concentrations, thousands of people were suddenly stricken with heart and respiratory problems, and scores died. Alarmed by such incidents, industrial countries enacted air quality laws designed to keep urban pollution at safer levels. While better protected from pollution's acute toxic effects, many people remain exposed to pollution levels that can cause insidious chronic effects.[60]

Just as scientists cannot unequivocally prove that air pollutants cause forest damage, health researchers rarely can prove that pollution causes specific cases of human illness and death. Factors such as occupational exposure to chemicals, cigarette smoking, and other health hazards confound the picture. Yet as an added stress on the human body, air pollutants reduce the productivity and lifespans of susceptible people, just as they do with trees. The U.S. Office of Technology Assessment estimates that the current mix of sulfates and particulates in ambient air may cause 50,000 premature deaths in the United States each year—about 2 percent of annual mortality. Especially vulnerable are the more than 16 million people already suffering from emphysema, asthma, and other chronic respiratory disorders.[61]

Ozone—the same pollutant damaging crops and trees—poses one of the most persistent air pollution problems in industrial countries. The maximum hourly ozone levels recommended by the World Health Organization are exceeded frequently in many large cities of Europe, Japan, and North America. As many as four out of every ten Americans are exposed to high ozone concentrations during the spring and summer, when weather conditions favor ozone's formation. Exposure to high levels of ozone, as with most other air pollutants, increases the number and severity of respiratory problems.[62]

In growing Third World cities, uncontrolled emissions from power plants, factories, and automobiles have added substantially to those from the burning of firewood and coal in homes. Between 1976 and 1980, annual sulfur dioxide concentrations in São Paulo, Brazil, averaged 25 percent higher than the U.S. standard set to protect human health. In the late seventies, Bombay and Calcutta were among the Indian cities registering levels of particulate matter far exceeding recommended limits. Similarly, in Beijing, China, sulfur dioxide concentrations for 1982 averaged eight times higher than the nation's primary standard; levels in the southwest city of Chongqing averaged 21 times higher. Health officials in Shanghai reportedly hold air pollution primarily responsible for a higher rate of lung cancer deaths there than elsewhere in the country.[63]

Heavy metals released to the environment during combustion, smelting, incineration, and other industrial processes have received much less attention than sulfur and nitrogen oxides and the other so-called conventional pollutants. Yet they are an increasing cause for concern.

Some metals—including copper, iron, and zinc—are essential nutrients needed by the body in small amounts. Others such as lead, cadmium, and mercury, serve no nutritional function. If introduced to the body in large enough quantities, either type can cause varying toxic effects, including cancer, and damage to the liver, kidneys, and central nervous system.[64]

Scientists have long known that high levels of lead in a person's blood can cause serious health damage. Several published papers on lead's toxic effects appeared in the mid-1800s. Since the 1920s, when petroleum refiners began adding lead to gasoline, people's exposure to this heavy metal has increased greatly. Over this same time, advances in biomedical technologies have caused health officials to lower the level of lead in blood considered safe. Lead is especially hazardous to children, since their bodies may absorb up to ten times more of a nonnutrient metal than an adult's will. An estimated 675,000 young children in the United States have high concentrations of lead in their blood. The effects vary with the quantities of lead present, but include damage to the kidney, liver, nervous system, and reproductive system; impaired growth; and interference with blood synthesis. With the burgeoning use of automobiles over the last few decades, millions of children worldwide have become exposed to potentially toxic amounts of lead.[65]

Risks from some metals—including cadmium, lead, and mercury—amplify with their ability to increase in concentration as they move up the food chain. Probably the most publicized disaster from such bioaccumulation occurred in the early fifties around Japan's Minamata Bay. There, the release of industrial mercury compounds accumulated in fish and shellfish. At least 700 people eating the bay's seafood acquired toxic levels of mercury, resulting in damage to their central nervous systems. In the United States, dangerous levels of mercury led the Wisconsin Department of Natural Resources to issue a warning in April 1985 against eating certain species of fish from 15 Wisconsin lakes. Similarly, in Poland's Gdansk Bay, quantities of mercury found in herring, cod, and flatfish have considerably exceeded permissible levels.[66]

In recent years, scientists have found that acid rain may magnify health risks from metals. Surveys of regions receiving acid precipitation, along with experiments in which lakes are purposely acidified, show that several metals—including aluminum, cadmium, mercury, and lead—become more soluble as acidification progresses. Acidic water can thus leach metals from soils and lake sediments into underground aquifers, streams, and reservoirs, potentially contaminating edible fish and water supplies. It can also dissolve toxic metals from the pipes and conduits of municipal or home water systems, contaminating drinking water.[67]

Studies in Sweden have shown that concentrations of zinc, copper, lead, and cadmium in acidic groundwater may be 10 to 100 times greater than background levels. So far, the measured concentrations do not exceed safe limits for drinking. In the United States, by contrast, drinking water samples in New York's Adirondack Mountains—a region receiving high rates of acid precipitation—have had lead concentrations up to 100 times higher than standards set to protect human health. At greatest risk are the millions of rural residents who draw water from private wells, since their drinking water is not routinely monitored for contaminants.[68]

Acid rain's ability to mobilize aluminum, the most abundant metallic element in the earth's crust, appears particularly disturbing. Aluminum normally remains bound up in soil minerals and is thus harmless to living organisms. Rendered soluble by increasing acidification, however, its concentration in lakes and streams has risen to levels harmful to aquatic life, and in some soils may be damaging to trees. Until the early seventies, the metal was thought to pose little or no hazard to human health. Then a strange syndrome that developed in kidney patients was traced to high concentrations of aluminum in the water supplies of some kidney dialysis centers. The large quantities of water used during dialysis treatment apparently allowed aluminum to accumulate to toxic levels in some patients' bodies, leading to death.[69]

Recently, some researchers have suggested a possible link between aluminum and Alzheimer's disease, a pervasive, degenerative disorder causing severe loss of memory and mental function. The association came to light from studies of the abnormal clumps of nerve-cell fibers—called neurofibrillary tangles—that characterize the brains of Alzheimer's victims. Using sensitive X-ray techniques, Dr. Daniel Perl, a neuropathologist at the University of Vermont, found significant accumulations of aluminum in the tangles, which were absent from normal control samples. Corroborating evidence comes from Perl's studies of native populations in Guam and two other locales of the western Pacific. All three populations have exhibited unusually high incidences of neurodegenerative disorders; all three regions have soils rich in bauxite, an ore of aluminum.[70]

Just how aluminum gets into the damaged nerve cells of Alzheimer's victims—and whether its presence is a cause or an effect—remains unknown. The destructive effects of aluminum on fish and trees suggest that these organisms have not adapted successfully to the altered chemistry of their environs. According to Perl, we can only speculate at this point as to whether humans will find themselves "in a similar vulnerable state."[71]

A third set of health hazards, along with fossil fuel pollutants and metals, stem from the introduction of synthetic chemicals to the environment. A chemist first prepared an organic compound without using a living organism more than 150 years ago. Over the last four decades, however, the number and types of chemicals created have proliferated almost beyond belief. The vast majority never leave the laboratory. Yet an estimated 70,000 chemicals presently are in everyday use, with between 500 and 1,000 new ones added to the list each year.[72]

By some measures, public fears raised about toxic chemicals appear incommensurate with the known risks these chemicals pose. Estimates of the share of cancer deaths caused by synthetic chemicals vary, but the most widely accepted range from 1 percent to as many as 10 percent. Compared to tobacco—which in the United States causes an estimated 30 percent of cancer deaths and nearly one-fifth of all deaths—known risks from synthetic chemicals pale in importance. Nonetheless, some investigators believe these compounds account for tens of thousands of deaths each year in the United States alone. Because of the long lag time—often 20 to 40 years—between exposure to a cancer-causing chemical and the appearance of the disease, the number of cancers induced by synthetic substances may increase markedly over the coming decades.[73]

Moreover, perhaps the greatest risks posed by manufactured chemicals derive not from what *is* known about their health effects, but rather from what is *not* known. So little data exists on the toxicity of chemicals now in use and on the extent of human exposure to them, that estimates of the total health risk they collectively present can only be educated guesses at best. The U.S. National Research Council (NRC) estimates that about 53,500 chemicals are used commercially in the United States. From available listings, chemicals fall into five categories: pesticides and pesticide ingredients, cosmetics, drugs, food additives, and a broad category of "chemicals in commerce" comprising compounds listed in the EPA's inventory of toxic substances. This latter category includes most industrial chemicals. They and the pesticides present the greatest threats to the general population through inadvertent exposure. Pesticides may leave residues in food, leach to underground water supplies, or spread through the air. Similarly, industrial chemicals may be released to the air, or when stored or disposed of on the land, may seep into drinking water.[74]

Despite the potential for widespread human exposure, most synthetic chemicals have received little or no testing for toxicity. No information on toxic effects is available for an estimated 79 percent of the chemicals in commerce. Less than a fifth have been tested for acute effects, and less than a tenth for chronic (e.g., cancer-causing), reproductive, or mutagenic effects. Moreover, the NRC found that chemicals produced in large volumes were tested no more frequently or thoroughly than those produced in smaller volumes.[75] Given how little is known about the extent of people's exposure to these substances, their introduction to the environment is akin to playing Russian roulette with human health.

Pesticides have generally received more extensive testing, but serious gaps remain. Charles Benbrook, executive director of the Board on

"Most synthetic chemicals have
received little or no testing for
toxicity."

Agriculture of the National Academy of Sciences, estimates that between 60 and 75 percent of the pesticides used on food and put on the market within the last decade have met the EPA's current standards for toxicity testing. Some products registered prior to the mid-seventies are now being evaluated, but hundreds of inadequately tested chemicals remain in use. Since these older pesticides typically cost less than the newer ones, they remain especially widely used by farmers in Third World countries. Moreover, following an EPA request for toxicity data, chemical manufacturers recently removed from the U.S. market at least 50 active pesticide ingredients, presumably because the required tests would suggest toxicity. While a promising sign that regulators can weed out hazardous substances, this response also suggests that millions of people may have been exposed to unsafe chemicals during recent decades. In countries where these compounds remain in use, the potential hazard persists.[76]

In the absence of adequate testing, knowledge of adverse chemical exposure may come only after serious health consequences arise. A classic example involved diethylstilbestrol (DES), which caused vaginal cancer in daughters of women given this compound during a critical period of their pregnancy. Men reportedly have suffered reproductive effects from occupational exposure to a variety of chemicals, including vinyl chloride, kepone, lead, and some common pesticides. Indeed, at least 20 chemicals have been associated with adverse reproductive effects in men or women, typically through exposure in the workplace. Harm to the general population is much more difficult to detect and prove.[77]

Recent findings of widespread human contamination, however, raise concerns that pervasive hazards may exist. Some investigators now believe, for example, that dioxins—among the most toxic chemicals known—are present in virtually everyone living in industrial countries. They may enter the food chain through use of the compound pentachlorophenol as a pesticide, through soil and water contamination around chemical manufacturing sites, and from the fallout of municipal and industrial incinerators burning certain organic chemicals. Analysts have found dioxins in such common foods as fish, eggs, chicken, and pork chops.[78]

Once ingested, dioxins are stored in the body's fat. The levels detected in the general population are far below that known to cause acute toxic effects in humans, but the long-term effects of low-level contamination remain unclear. Dioxins have produced tumors in animals and have been linked with certain rare cancers in people. Recent evidence also suggests that dioxin may damage the immune system, which would weaken people's ability to fight disease. Breast-feeding babies appear especially at risk from any toxic effects. Researchers have calculated that through the dioxin-contaminated fat in breast milk, an infant nursed for one year could acquire 18 times the "allowable" lifetime exposure estimated by the federal Centers for Disease Control.[79]

Additional evidence of widespread chemical hazards comes from research at Florida State University that points to environmental chemicals as contributing to a decline in male fertility. In U.S. males, sperm density—a measure of fertility—has apparently diminished significantly since mid-century. A 1979 study of 132 university students showed that nearly one out of four men had sperm densities low enough to reduce their reproductive success. Each sample of seminal fluid contained synthetic chemicals, including pentachlorophenol—widely used as a wood preservative as well as a pesticide—polychlorobiphenyls (PCBs), and metabolites of DDT. According to the researchers, toxic substances accounted for more than a quarter of the variation in sperm density found among the students.[80]

Chemicals that diminish sperm density also stand a good chance of being mutagenic, carcinogenic, or both. Chemically reduced sperm counts may be caused by DNA damage that slows the process of cell division involved in sperm creation. Damage to DNA is believed to be a crucial step in the development of cancer. Interestingly, since 1950, testicular cancer rates have doubled among white males and tripled among black males. Michael Castleman, managing editor of *Medical Self-Care*, writes that "a century ago testicular cancer was virtually

"Virtually all chemicals present in
a mother's blood will get into her milk,
which an infant may then ingest."

unheard of in men under 50...; today it is one of the most common cancers in men between the ages of 15 and 34."[81]

Even if scientists could thoroughly test each of the tens of thousands **41** of chemicals in use, uncertainties would remain about their effects on human health. Whether through drinking water, the air, or food, people typically are exposed to more than one chemical at a time. Combinations of chemicals can pose different degrees of hazard than the same chemicals in isolation. The body's primary defense against foreign agents is a chemical network known as the microsomal enzyme system, or MES. It evolved to regulate the amounts of certain hormones, vitamins, and other substances in the body, and works to detoxify chemicals or prepare them for excretion. The presence of a foreign substance usually induces MES activity. However, many environmental chemicals—including some pesticides—inhibit the MES. Moreover, tests have shown that more than 50 environmental chemicals yield metabolites in the body that are more toxic than the parent chemical itself.[82]

Arthur J. Vander, a professor of physiology at the University of Michigan, points out that when considering how the body metabolizes environmental chemicals, "we must never assume that children are simply little adults." The MES is not fully mature in newborn babies and very young infants. Babies may have limited access to the environment, but they are exposed to environmental chemicals through their mothers. Virtually all chemicals present in a mother's blood will get into her milk, which an infant may then ingest. Also, since the placenta that nourishes a developing fetus is designed for diffusion, chemicals in a mother's bloodstream may pass, in varying amounts, into her baby. The concentration of DDT in fetal blood, for example, is typically almost half that in maternal blood.[83]

With advances in medical technologies and more sophisticated epidemiological studies, evidence linking chemicals to adverse health effects seems likely to grow. Researchers are now investigating the possibility, for example, that Parkinson's disease is associated with exposure to environmental chemicals. The disease usually develops late in life, is of unknown cause, and involves the degeneration of neurons in certain portions of the brain. Its victims suffer tremors and partial or complete paralysis. A team of researchers in Montreal, Canada, has found a very strong correlation between the incidence of Parkinson's disease and the level of pesticide use in nine regions of Quebec Province. Quebec's principal agricultural region had the highest rural incidence of Parkinson's—0.89 per thousand people, compared with 0.13 per thousand in areas with little pesticide use. Paraquat-like pesticides closely resemble a chemical known to induce a parkinsonian-like state in humans and animals. While controversial and by no means proven, the suggested link between Parkinson's disease and common chemicals raises some profound implications both for human health and society's use of synthetic substances.[84]

Just how extensive and serious the health consequences of chemical exposure may become is impossible to judge. More research is needed to unravel the complex and sometimes subtle ways in which chemicals affect the human body. Since many toxic effects appear several decades after exposure to the offending chemical, the full implications of the chemical age will take time to realize. Given the thousands of chemicals introduced into the environment and the lack of knowledge about their health effects, some unpleasant surprises may lie in store.

Minimizing Risks

Industrial societies have spawned multiple and rapidly increasing changes in the earth's chemistry. The resulting threats to food security, forests, and human health pose substantial risks over the coming decades. Unfortunately, powerful barriers exist to effectively limiting these threats: Much scientific uncertainty surrounds them; their full consequences may not appear for several decades; and many of their future economic and social costs remain unquantifiable. Taking no action, however, invites potentially irreversible and disastrous effects. A strategy of minimizing risks—where possible, through measures that remedy several problems simultaneously—deserves immediate support.

Because of society's past and present dependence on fossil fuels, a change in the world's climate is already inevitable. Yet since carbon dioxide is the key variable in the climate equation, the magnitude of climatic change—and the pace at which it unfolds—will depend greatly on the future use of coal, oil, and natural gas. If worldwide carbon emissions from fossil fuels return to their pre-1973 rate of growth—more than 4 percent per year—the atmospheric concentration of CO_2 will reach double preindustrial levels in about 40 years. (See Table 6.) On the other hand, holding that growth to 1 percent per year would delay a CO_2 doubling for more than a century.

A goal of limiting the annual growth of carbon emissions to 1 percent may have seemed utterly unrealistic ten years ago. Now, however, it appears entirely feasible. For the decade following 1973, worldwide carbon emissions grew at an encouragingly low average rate of 1.1 percent per year, just one-quarter the pre-1973 rate. Carbon emissions actually fell for four consecutive years, 1980 through 1983. Yet preliminary data for 1984 show an increase over 1983 emissions. This suggests that an upward trend could resume, a possibility enhanced

Table 6: Projected Dates for a Doubling of CO_2 over Preindustrial Levels for Different Rates of Growth in Fossil Fuel Emissions

Annual Worldwide Growth in Fossil Fuel Emissions	Projected CO_2 Doubling Time	
(percent)	(year)	(years from present)
4	2026	40
3	2036	50
2	2054	68
1	@2100	@114

Source: Adapted from William W. Kellogg and Robert Schware, *Climate Change and Society* (Boulder, Colo.: Westview Press, 1981).

by the recent drop in oil prices. Maintaining a 1 percent worldwide rate of growth will require a virtual cap on carbon emissions from industrial countries to allow for needed growth in energy use in the Third World.[85]

At the October 1985 gathering in Villach, Austria, scientists from 29 nations concurred that the rate and degree of future warming could be "profoundly affected" by government policies that curb the use of fossil fuels. No nation, however, has yet taken steps explicitly geared toward this end. Reducing fossil fuel combustion would also reduce acid rain and air pollution, thereby relieving threats to forests and to human health. Indeed, without West Germany's 8 percent decline in total energy consumption between 1979 and 1984, air pollution damage to the nation's forests might have progressed even further than it has. By focusing only on pollution control technologies—such as scrubbers for power plants and catalytic converters for automobiles—virtually all nations are neglecting opportunities to limit acid-forming pollutants and the CO_2-buildup simultaneously.[86]

Commitments to increase energy efficiency, to recycle more materials, and to meet new energy needs from sources other than fossil fuels, could cost-effectively reduce fossil fuel emissions and all their associated risks. The rise in energy costs during the seventies triggered some impressive efficiency gains, but vast potential remains for cutting energy use in industries, automobiles, and homes. In Sweden, houses with heating requirements one-seventh that of an average Swedish home are now on the market. Legislation there calls for a 30 percent energy savings in buildings over ten years, which if achieved, will cut the nation's total energy consumption by 15 percent. Federal and state governments in West Germany jointly provided nearly $2 billion between 1978 and 1982 to promote efforts to reduce energy used for heating. The program continues to support new technologies, such as heat pumps, solar designs, and hook-ups to district heating systems. Projections indicate a 30 percent drop in the amount of energy needed for space heating by the year 2000. Because of these and other efforts to boost efficiency, West German analysts now expect no increase in the nation's total primary energy needs through the year 2000.[87]

"Virtually all nations are neglecting
opportunities to limit acid-forming
pollutants and the CO_2-buildup
simultaneously."

Setting standards for residential appliances, which in the United States account for about a third of all electricity needs, could also cut energy use substantially. In the absence of federal initiatives, several states have taken the lead. California's standards call for a 50 percent increase in the efficiency of refrigerators and freezers by 1992, which would eliminate the need for one large power plant. Prospects for federal standards improved in the summer of 1985 when a federal appeals court ordered the Department of Energy to reappraise standards for common appliances. Analyses show that even conservative standards—ones ensuring a payback of any added consumer costs within the time period allotted for credit loan repayments—would reduce the nation's energy needs in the year 2000 by 10 percent. Such savings, which do not include those induced by market forces, would eliminate the need for 15 large power plants. More stringent standards, of course, could achieve much more.[88]

45

Stepping up recycling efforts for aluminum, glass, paper, and other materials would further reduce energy needs and power plant emissions, as well as pollution directly resulting from materials manufacturing. Making paper from recycled wastepaper rather than from new wood cuts energy use by a third to half and air pollutants by as much as 95 percent. Producing a ton of aluminum from virgin bauxite ore takes the energy equivalent of 8.1 metric tons of coal. Producing the same amount from recycled scrap takes the equivalent of 0.4 tons, 95 percent less. Worldwide recycling rates are far below their potentials—less than one-third for aluminum and about one-fourth for paper. By eliminating waste, recycling also reduces the need for landfilling and incineration, lessening the corresponding risks of groundwater and air pollution.[89]

Shifting energy sources away from those that emit large quantities of carbon, sulfur, nitrogen, and hydrocarbon compounds can also help reduce simultaneously the costs of air pollution, acid rain, and climate change. Selected use of natural gas in place of coal and oil could reduce pollution fairly quickly and economically, and provide a practical bridge to an energy economy less dependent on all fossil fuels. Per unit of useful energy output, natural gas emits 30 percent less carbon than oil, and 42 percent less than coal. (See Table 7.) It releases

Table 7: Carbon Emissions per Unit of Energy Produced from Fossil Fuels

Fuel	Carbon Emissions	Emissions as Percent of those from Coal
	(kilograms/gigajoule[1])	(percent)
Natural Gas	13.8	58
Oil	19.7	82
Coal	23.9	100
Synthetic Fuels[2]	35.8	150
Coal-fired Electricity	68.3	286

[1]One gigajoule equals one billion joules or 949,000 Btu. [2]Oil and gas.

Source: John A. Laurmann, "The Role of Gas in the Emission of Carbon Dioxide and Future Climate Change," *Gas Research Insights*, July 1985.

fewer nitrogen oxides than either coal or oil, and virtually no sulfur. On the other hand, use of synthetic fuels, which some see as an eventual replacement for dwindling oil supplies, results in a substantial increase in carbon emissions compared with any of the natural fossil fuels.[90]

A few nations have begun to shift away from fossil fuels. In West Germany, alternative sources are expected to produce 17 percent of primary energy in the year 2000, up from 11 percent at present and 4 percent in 1960. Nuclear power dominates this non-fossil-fuel contribution. Yet where concerns about radioactive waste disposal, high costs, or safety lead energy planners to shy away from the nuclear option, other sources—including solar, wind, and hydro power—could greatly expand the share of energy generated without fossil fuels. In the United States, nearly 1,300 small-scale power projects using a mix of alternative sources have been planned since 1980. Their combined capacity will equal that of 25 large coal or nuclear plants. Sweden's energy policy includes reducing dependence on oil,

"A commitment among industrial
countries to cap emissions of carbon
now appears essential."

while also phasing out nuclear power. Along with increasing energy efficiency, Sweden plans to meet a greater share of its needs from wind, water, and solar power.[91]

Achieving meaningful reductions in fossil fuel emissions will require concerted efforts by many nations. One promising sign of cooperation is the commitment of 21 countries to reduce emissions of sulfur dioxide by at least 30 percent of 1980 levels by 1993. (Unfortunately, the United States and Great Britain—North America and Western Europe's largest SO_2 emitters—are not among them.) Yet a commitment among industrial countries to cap emissions of carbon, while substantially reducing those of sulfur and nitrogen oxides, now appears essential to minimize the risks that have arisen from fossil fuel combustion.[92]

While measures to curb the use of fossil fuels would reduce risks most broadly, at least three other actions relating to specific threats appear warranted. The synthetic chlorofluorocarbons damage the earth's protective layer of ozone and also contribute greatly to climatic change. Lessening these threats requires substantial reductions in CFC emissions, and a worldwide ban on nonessential aerosol uses of CFCs would be a cost-effective first step. A few nations have already restricted or banned such uses. This action actually proved beneficial to the U.S. economy, with readily available substitutes saving consumers an estimated $165 million in 1983 alone. Under the auspices of the United Nations Environment Program, international negotiations regarding CFCs are in progress, but so far have resulted only in a framework for adopting control measures if they are deemed necessary.[93]

Preserving forests, essential for a host of reasons, plays a dual role in minimizing risks of climate change. The clearing and burning of tropical forests contributes substantially to the annual release of carbon to the atmosphere. Trees also remove carbon dioxide from the air during photosynthesis. Protecting natural forests and planting trees can thus do much to minimize the threat of climate change. In mid-1985, a promising development emerged with the unveiling of an ambitious tropical forest protection plan. Designed by an international task force coordinated by the Washington, D.C.-based World Resources Institute, and supported by leading aid agencies, it calls for investments totalling $8 billion over five years in tree-planting projects and efforts to arrest deforestation. If adequately funded and implemented, the plan may help begin a much-needed reversal of deforestation trends.[94]

Threats to health from the increased presence of lead in the environment could be reduced cost-effectively by phasing out the use of lead in gasoline. Studies in the United States suggest that gasoline accounts on average for about half of the lead in people's blood. In 1973, the United States initiated regulations controlling lead in fuel, and they have effectively reduced the amount of lead in the environment. Measurements show that between the mid-seventies and early eighties the amount of pollutant lead carried to the Gulf of Mexico by the Mississippi River—which drains 40 percent of the contiguous United States—declined by 40 percent.[95]

Through a detailed cost-benefit analysis, the EPA determined in 1985 that a further reduction in the allowable lead content of gasoline from 1.1 grams per gallon to 0.1 would yield health benefits far exceeding the added costs petroleum refiners would incur. The agency estimated net benefits of at least $1.3 billion (1983 dollars) for 1986 alone. That new lower limit is now in effect. Most countries, however, lag behind the United States in controlling lead in gasoline. Many European nations are just introducing unleaded fuel, and standards are minimal or nonexistent in most Third World countries. Besides reducing health risks, use of unleaded fuel is essential for catalytic converters, the state-of-the-art technology for controlling the carbon and nitrogen compounds emitted by automobiles.[96]

Lessening the risks posed by synthetic chemicals may require a fundamental rethinking of the way these substances are commercialized, used, and discarded. Setting priorities and increasing funding for more extensive toxicity testing is urgently needed. Thorough testing of a chemical using mice or rats can take as long as five years and cost up to $500,000. At such high costs, little hope exists of fully testing the tens of thousands of chemicals already on the market. Adequately testing the 500 to 1,000 new chemicals created each year for use in the United States could require as much as $500 million annually.[97]

Short-term tests costing only a few hundred dollars offer a useful screening mechanism for setting priorities for additional testing. Adequately protecting the public, however, will require that industries profiting from the sale of chemicals take more responsibility for ensuring chemical safety. Under the amended U.S. law regulating pesticides, manufacturers must show that new products do not pose unacceptable health risks. Laws governing other categories of chemicals, however, place the burden of proof on government regulators. In such cases, a public agency must demonstrate that a chemical poses an unacceptable risk before taking action to restrict or ban it. Besides creating backlogs and requiring large expenditures of tax dollars, such a policy can allow harmful chemicals to remain in use for many years. J. Clarence Davies of the Conservation Foundation points out that additional toxicity testing, though expensive, in most cases amounts to a small share of the total cost of producing a chemical.[98]

Introducing chemicals to the environment without knowledge of their health effects implicitly presumes that the benefits of these chemicals outweigh their costs. In addition to the fact that these costs are rarely known, this presumption does not always hold. Since mid-century, for example, farmers have greatly increased their dependence on chemicals to kill insects and other pests. Yet the share of crops lost to pest damage has diminished little, if at all. Besides killing pests, chemicals often kill the pests' predators, thereby eliminating effective natural controls. Furthermore, pests rapidly become resistant to the chemicals created to kill them, placing farmers on a costly "pesticide treadmill."[99]

More effective use of biological pest controls could greatly reduce agriculture's dependence on chemicals and the associated threats to health. Over the last two decades, researchers have developed techniques collectively known as "integrated pest management" (IPM), which seeks to reduce pest damage while minimizing pesticide use. It may include, for example, introducing natural predators to prey on pests, monitoring pest population levels and applying chemicals only when necessary, or applying pesticides at the most vulnerable point of a pest's life cycle. By reducing chemical costs, IPM usually benefits farmers economically. In the Texas High Plains, for example, IPM programs to control the cotton boll weevil increased farmers' annual net profits by $27 million.[100]

In the United States, IPM efforts over the last 10 to 15 years have been directed intensively at cotton, grain sorghum, and peanuts. For each of these crops, insecticide use has declined markedly. Between 1971 and 1982, the amount of insecticide applied per hectare fell by 40 percent for sorghum, 75 percent for cotton, and 81 percent for peanuts. (See Table 8.) In contrast, insecticide use per hectare of corn and soybeans—crops that did not receive intensive IPM efforts—slightly increased. Support for research and development of IPM in recent years has not been commensurate with its potential benefits to agriculture and the environment.[101]

Incentives are also needed to reduce the volume and toxicity of hazardous chemicals requiring disposal. Many countries would benefit from Europe's experience in this area. Detoxification of wastes is standard procedure in at least a half-dozen European countries. The Netherlands' Chemical Waste Act of 1976 prohibited land disposal of many toxic substances, spurring the development of a thriving waste detoxification industry. In Denmark, a large centrally located facility—the Kommunekemi plant in Nyborg—accepts wastes from any community in the country that cannot treat or safely dispose of them. Twenty major stations around the country collect toxic chemicals for transport to Nyborg. The plant destroys or recycles the waste, and if that proves impossible, the waste is stored until technologies to treat it are developed. West Germany's approach is similar, but more decentralized: 85 percent of its hazardous waste is sent to 15 large treatment plants for detoxification or recycling.[102]

Table 8: Effects of Integrated Pest Management on Insecticide Use, United States, 1971-1982

Crop	Use of IPM	Insecticide Use 1971	1982	Change
		(kilograms/hectare)		(percent)
Corn	minimal	0.38	0.41	+ 8
Soybeans	minimal	0.15	0.17	+13
Grain Sorghum	intensive	0.30	0.18	−40
Cotton	intensive	6.63	1.68	−75
Peanuts	intensive	4.48	0.86	−81

Source: R. E. Frisbie and P. L. Adkisson, "IPM: Definitions and Current Status in U.S. Agriculture," in Marjorie A. Hoy and Donald C. Herzog, eds., *Biological Control in Agricultural IPM Systems* (Orlando, Fla.: Academic Press, Inc., 1985).

In contrast to these strategies designed to keep toxics out of the environment, 80 percent of the hazardous waste generated in the United States is dumped on the land. Land disposal typically costs at least 20 to 30 percent less per ton of waste than safer alternatives. To narrow this cost differential and generate funds for waste management, several states—including California and New York—have begun to tax toxic waste disposal. Yet a fee levied at the federal level could more readily and effectively move the nation away from the hazards of dumping, while providing much-needed funding for cleaning up abandoned hazardous waste sites.[103]

No nation acting alone can avert the costly consequences of altering the earth's chemistry. Air pollutants and acid rain waft easily across political boundaries. Carbon dioxide emissions anywhere contribute to climate change everywhere. Pesticides produced in one country may freely be traded for use in others. Yet translating shared risks

52

into cooperation aimed at minimizing them is no easy task. A decision to place a hefty tax on fossil fuel combustion, for example, would have serious political and economic repercussions. Few nations would view it in their interest to adopt such a preventive measure without guarantees that others will do likewise.

Attaining the sustained cooperation needed to deal with global environmental change will be an arduous process, one that cannot begin too soon. Strong international ties within the scientific community lay a helpful foundation. A proposed "International Geosphere-Biosphere Program" directed by the International Council of Scientific Unions would engage scientists from many nations in a coordinated effort to further unravel the nature and magnitude of global change, and humanity's role in it. The program may begin around 1990, and would likely last at least ten years. Yet political action need not and should not await more conclusive scientific results. At the October 1985 conference in Villach, Austria, scientists from 29 nations agreed that the prospects of climatic change are understood sufficiently that "scientists and policy-makers should begin an active collaboration" to explore policy options.[104]

Many different institutions can help build the cooperation needed among governments. The United Nations Environment Program, the U.N. Economic Commission for Europe, the European Economic Community, the World Meteorological Organization, and others have in various ways been instrumental in achieving progress toward global environmental management. But only with leadership from individual nations will concrete measures result. Sweden's efforts to make acid rain a top priority on the international environmental agenda, and West Germany's call for stricter pollution controls on European power plants and automobiles are just two examples of the kind of leadership needed. Action by just a few countries can lead to action by many. Ten nations initially made the commitment in March 1984 to reduce their sulfur dioxide emissions by 30 percent within a decade; at present, 21 nations have so committed themselves. Meaningful reductions in worldwide carbon emissions would begin with concerted measures by just three nations—China, the Soviet Union, and the United States, the world's three largest users of coal.

Fossil fuels and chemicals have figured prominently in society's pursuit of economic growth and higher standards of living. Yet changes in the earth's chemistry wrought by their use threaten the integrity of natural systems upon which future growth and human well-being depend. Alternatives to the present course exist. By failing to act, we thrust upon ourselves and the next generation potential crises we have the capacity to avert.

Notes

1. J.E. Lovelock, *Gaia: A New Look at Life on Earth* (New York: Oxford University Press, 1979).

2. Total cost estimate from John A. Laurmann, "Strategic Issues and the CO_2 Environmental Problem," reprint from W. Bach et al., eds., *Carbon Dioxide: Current Views and Developments in Energy/Climate Research* (Dordrecht, The Netherlands: D. Reidel Publishing Company, 1983).

55

3. H.J. Ewers et al., "Zur Monetarisierung der Waldschäden in der Bundesrepublik Deutschland," paper presented at Symposium on Costs of Environmental Pollution, Bonn, West Germany, September 12-13, 1985.

4. U.S. Office of Technology Assessment (OTA), *Acid Rain and Transported Air Pollutants: Implications for Public Policy* (Washington, D.C.: U.S. Government Printing Office, 1984).

5. G. Tyler Miller, Jr., *Living in the Environment: Concepts, Problems, and Alternatives* (Belmont, Calif.: Wadsworth Publishing Company, Inc., 1975).

6. A.M. Solomon et al., "The Global Cycle of Carbon," R.M. Rotty and C.D. Masters, "Carbon Dioxide from Fossil Fuel Combustion: Trends, Resources, and Technological Implications," and R.A. Houghton et al., "Carbon Dioxide Exchange Between the Atmosphere and Terrestrial Ecosystems," in John R. Trabalka et al., *Atmospheric Carbon Dioxide and the Global Carbon Cycle* (Washington, D.C.: U.S. Government Printing Office, 1985); see also G.M. Woodwell et al., "Global Deforestation: Contribution to Atmospheric Carbon Dioxide," *Science*, December 9, 1983; R.A. Houghton et al., "Net Flux of Carbon Dioxide from Tropical Forests in 1980," *Nature*, August 15, 1985; data for figure provided by Ralph M. Rotty, Institute for Energy Analysis, Oak Ridge Associated Universities, Oak Ridge, Tenn., May 1986.

7. Roger Revelle, "Carbon Dioxide and World Climate," *Scientific American*, August 1982; Roger Revelle, "The Oceans and the Carbon Dioxide Problem," *Oceans*, Summer 1983; preindustrial concentration from Eric W. Wolff and David A. Peel, "The Record of Global Pollution in Polar Snow and Ice," *Nature*, February 14, 1985.

8. Revelle, "Carbon Dioxide and World Climate"; Carbon Dioxide Assessment Committee, National Research Council (NRC), *Changing Climate* (Washington, D.C.: National Academy Press, 1983); Stephen Seidel and Dale Keyes, *Can We Delay a Greenhouse Warming?* (Washington, D.C.: U.S. Environmental Protection Agency, 1983). Ice Age comparison from National

Aeronautics and Space Administration (NASA), Goddard Space Flight Center, "Potential Climatic Impacts of Increasing Atmospheric CO_2 With Emphasis on Water Availability and Hydrology in the United States," prepared for U.S. Environmental Protection Agency (EPA), Washington, D.C., 1984.

9. NRC, *Global Change in the Geosphere-Biosphere* (Washington, D.C.: National Academy Press, 1986); sources from D.H. Ehhalt, "Methane in the Global Atmosphere," *Environment*, December 1985; estimates of warming from Gordon J. MacDonald, "Climate Change and Acid Rain," The MITRE Corporation, McLean, Va., December 1985; and V. Ramanathan et al., "Trace Gas Trends and their Potential Role in Climate Change," *Journal of Geophysical Research*, June 20, 1985.

10. Global estimates from P.J. Crutzen and M.O. Andreae, "Atmospheric Chemistry," in T.F. Malone and J.G. Roederer, eds., *Global Change* (Cambridge, Great Britain: Cambridge University Press, 1985); data for figure from EPA, *National Air Pollutant Emission Estimates, 1940-1984* (Research Triangle Park, N.C.: 1986).

11. MacDonald, "Climate Change and Acid Rain"; Ramanathan et al., "Trace Gas Trends and their Potential Role in Climate Change."

12. Crutzen and Andreae, "Atmospheric Chemistry."

13. Swedish Ministry of Agriculture, *Proceedings: The 1982 Stockholm Conference on Acidification of the Environment* (Stockholm: 1982); James N. Galloway and Douglas M. Whelpdale, "An Atmospheric Sulfur Budget for Eastern North America," *Atmospheric Environment*, Vol. 14, 1980.

14. Walter W. Heck et al., "A Reassessment of Crop Loss from Ozone," *Environmental Science & Technology*, Vol. 17, No. 12, 1983; Environmental Resources Limited, *Acid Rain: A Review of the Phenomenon in the EEC and Europe* (London: Graham & Trotman Limited, 1983).

15. Jack G. Calvert et al., "Chemical Mechanisms of Acid Generation in the Troposphere," *Nature*, September 5, 1985; Ellis B. Cowling, "Acid Precipitation in Historical Perspective," *Environmental Science & Technology*, Vol. 16, No. 2, 1982.

16. Gene E. Likens et al., "Acid Rain," *Scientific American*, October 1979; Galloway and Whelpdale, "An Atmospheric Sulfur Budget for Eastern North America"; U.S. Interagency Task Force on Acid Deposition, National Acid

Precipitation Assessment Program, *Annual Report 1984* (Washington, D.C.: 1984).

17. James N. Galloway et al., "Trace Metals in Atmospheric Deposition: A Review and Assessment," *Atmospheric Environment*, Vol. 16, No.7, 1982; John H. Trefry et al., "A Decline in Lead Transport by the Mississippi River," *Science*, October 25, 1985.

18. A.H. Johnson et al., "Spatial and Temporal Patterns of Lead Accumulation in the Forest Floor in the Northeastern United States," *Journal of Environmental Quality*, Vol. 11, No. 4, 1982; Taiwan reference from "International Researchers Discuss Lead Pollution," *World Environment Report*, July 10, 1985; quote from H. Heinrichs and R. Mayer, "The Role of Forest Vegetation in the Biogeochemical Cycle of Heavy Metals," *Journal of Environmental Quality*, Vol. 9, No. 1, 1980.

19. Galloway et al., "Trace Metals in Atmospheric Deposition."

20. NRC, *Causes and Effects of Changes in Stratospheric Ozone: Update 1983* (Washington, D.C.: National Academy Press, 1984).

21. NRC, *Causes and Effects of Changes in Stratospheric Ozone*; Peter H. Sand, "The Vienna Convention is Adopted," *Environment*, June 1985; recent production increase from Chemical Manufacturers Association, "Production and Release of Chlorofluorocarbons 11 and 12," Washington, D.C., October 1985.

22. NASA, "Present State of Knowledge of the Upper Atmosphere," draft, January 1986; EPA, "Analysis of Strategies for Protecting the Ozone Layer," prepared for Working Group Meeting, Geneva, Switzerland, January 1985; NRC, *Causes and Effects of Changes in Stratospheric Ozone*; see also Paul Brodeur, "Annals of Chemistry," *The New Yorker*, June 9, 1986.

23. Ramanathan et al., "Trace Gas Trends and Their Potential Role in Climate Change"; "An Assessment of the Role of Carbon Dioxide and of Other Greenhouse Gases in Climate Variations and Associated Impacts," statement from conference cosponsored by United Nations Environment Program, World Meteorological Organization, and International Council of Scientific Unions, Villach, Austria, October 1985.

24. Rachel Carson, *Silent Spring* (Greenwich, Conn.: Fawcett Publications, Inc., 1962).

58

25. Miller, *Living in the Environment;* NRC, *Testing for Effects of Chemicals on Ecosystems* (Washington, D.C.: National Academy Press, 1981); David Pimentel and Clive A. Edwards, "Pesticides and Ecosystems," *BioScience,* July/August, 1982.

26. NRC, *Testing for Effects of Chemicals on Ecosystems;* Robert M. Garrels et al., *Chemical Cycles and the Global Environment: Assessing Human Influences* (Los Altos, Calif.: William Kaufmann, Inc., 1975); Arthur J. Vander, *Nutrition, Stress, and Toxic Chemicals* (Ann Arbor: University of Michigan Press, 1981); Lawrie Mott and Martha Broad, "Pesticides in Food: What the Public Needs to Know," Natural Resources Defense Council, Inc., San Francisco, Calif., March 1984.

27. Walter Orr Roberts, "It is Time to Prepare for Global Climate Changes," *Letter,* The Conservation Foundation, April 1983.

28. For a comprehensive review of climate issues see, NRC, *Changing Climate.*

29. Michael E. Schlesinger and John F.B. Mitchell, "Model Projections of the Equilibrium Climatic Response to Increased Carbon Dioxide," in Michael C. MacCracken and Frederick M. Luther, eds., *The Potential Climatic Effects of Increasing Carbon Dioxide* (Washington, D.C.: U.S. Department of Energy, 1985).

30. S. Manabe and R.T. Wetherald, "Reduction in Summer Soil Wetness Induced by an Increase in Atmospheric Carbon Dioxide," *Science,* May 2, 1986; Villach conference statement, "An Assessment of the Role of Greenhouse Gases in Climate Variations"; Dean Abrahamson, "Responses to Greenhouse Gas Induced Climate Change," testimony presented before the U.S. Senate Subcommittee on Toxic Substances and Environmental Oversight, Washington, D.C., December 10, 1985; William W. Kellogg, "Impacts of a CO_2-Induced Climate Change," reprint from Bach et al., eds., *Carbon Dioxide: Current Views and Developments.*

31. Revelle, "Carbon Dioxide and World Climate"; William W. Kellogg and Robert Schware, "Society, Science and Climate Change," *Foreign Affairs,* Summer 1982; Syukuro Manabe, Geophysical Fluid Dynamics Laboratory, Princeton, N.J., private communication, January 10, 1986.

32. Dean Abrahamson and Peter Ciborowski, "North American Agriculture and the Greenhouse Problem," Hubert H. Humphrey Institute of Public Affairs, University of Minnesota, Minneapolis, Minn., April 1983; Kellogg and Schware, "Society, Science and Climate Change."

33. M. Barth and J. Titus, eds., *Greenhouse Effect and Sea Level Rise: A Challenge for This Generation* (New York: Van Nostrand Reinhold Co., 1984); Villach conference statement, "An Assessment of the Role of Greenhouse Gases in Climate Variations"; Erik Eckholm, "Significant Rise in Sea Level Now Seems Certain," *The New York Times*, February 18, 1986.

34. Sylvan H. Wittwer, "Carbon Dioxide and Climate Change: An Agricultural Perspective," *Journal of Soil and Water Conservation*, May/June 1980; Sylvan H. Wittwer, "Rising Atmospheric CO_2 and Crop Productivity," *Hortscience*, Vol. 18, October 1983; Paul E. Waggoner, "Agriculture and a Climate Changed by More Carbon Dioxide," in NRC, *Changing Climate*.

35. Waggoner, "Agriculture and a Climate Changed by More Carbon Dioxide"; Sylvan H. Wittwer, "Carbon Dioxide Levels in the Biosphere: Effects on Plant Productivity," *CRC Critical Reviews in Plant Sciences*, Vol. 2, Issue 3; see also C. Mlot, "Trickle-Down Effects of Carbon Dioxide Rise," *Science News*, November 17, 1984.

36. Irrigation figures from W.R. Rangeley, "Irrigation and Drainage in the World," paper presented at the International Conference on Food and Water, Texas A&M University, College Station, Tex., May 26-30, 1985.

37. Roger R. Revelle and Paul E. Waggoner, "Effects of a Carbon Dioxide-Induced Climatic Change on Water Supplies in the Western United States," in NRC, *Changing Climate*.

38. Trend in irrigated area from U.S. Department of Agriculture, *Agricultural Statistics 1983*, and U.S. Department of Commerce, "Census of Agriculture"; present overconsumption in Lower Colorado from U.S. Geological Survey, *National Water Summary 1983—Hydrologic Events and Issues* (Washington, D.C.: U.S. Government Printing Office, 1984); irrigated area calculation assumes an annual consumptive demand of 5,500 cubic meters per hectare, which is 55 percent of the average per-hectare-withdrawals for irrigation estimated in Revelle and Waggoner, "Effects of a Carbon Dioxide-Induced Climate Change"; existing irrigated area also from Revelle and Waggoner.

39. Calculation based on average investment needs for new large-scale irrigation projects in the Third World of $5,000 per hectare, from Rangeley, "Irrigation and Drainage in the World."

40. Cost estimate from John A. Laurmann, "Strategic Issues and the CO_2 Environmental Problem"; William W. Kellogg, "The Socio-Economic Response: Human Factors in Environmental Change," paper presented at the

41. OTA, *Acid Rain and Transported Air Pollutants*.

42. Ibid.; Richard M. Adams et al., *The Economic Effects of Ozone on Agriculture* (Washington, D.C.: U.S. Government Printing Office, 1984).

43. Environmental Resources Limited, *Acid Rain: A Review of the Phenomenon in the EEC and Europe*; Maria Elena Hurtado, "A Hard Rain Begins to Fall...," *South*, November 1985.

44. Der Bundesminister Für Ernährung, Landwirtschaft und Forsten, "Neuartige Waldschäden in der Bundesrepublik Deutschland," Bonn, West Germany, October 1983; Federal Ministry of Food, Agriculture and Forestry, "1984 Forest Damage Survey," Bonn, West Germany, October 1984.

45. *Allgemeine Forstzeitschrift*, Munich, West Germany, November 11, 1985; G.H.M. Krause et al., "Forest Decline in Europe: Possible Causes and Etiology," paper presented at the International Symposium on Acid Precipitation, Ontario, Canada, September 1985; see also Susan Tifft, "Requiem for the Forest," *Time* (international edition), September 16, 1985.

46. Krause et al., "Forest Decline in Europe"; Dieter Deumling, Wissen, West Germany, private communication, March 1986; "Swiss Forests are Depleted Further by Pollution," *The New York Times*, December 9, 1985.

47. Thomas G. Siccama et al., "Decline of Red Spruce in the Green Mountains of Vermont," *Bulletin of the Torrey Botanical Club*, April/June 1982; Arthur H. Johnson and Thomas G. Siccama, "Acid Deposition and Forest Decline," *Environmental Science & Technology*, Vol. 17, No. 7, 1983.

48. Raymond M. Sheffield et al., "Pine Growth Reductions in the Southeast," Southeastern Forest Experiment Station, Asheville, N.C., November 1985; Arthur H. Johnson, "Assessing the Effects of Acid Rain on Forests of the Eastern U.S.," testimony before the U.S. Senate, Committee on Environment and Public Works, Hearings, February 7, 1984.

49. Sandra Postel, *Air Pollution, Acid Rain, and the Future of Forests* (Washington, D.C.: Worldwatch Institute, 1984); Krause et al., "Forest Decline in Europe"; L.W. Blank, "A New Type of Forest Decline in Germany," *Nature*, March 28, 1985; Bengt Nihlgård, "The Ammonium Hypothesis—An Additional Explanation to the Forest Dieback in Europe," *Ambio*, Vol. 14, No. 1, 1985.

50. B. Prinz et al., "Responses of German Forests in Recent Years: Cause for Concern Elsewhere?" and George H. Tomlinson, "Acid Deposition, Nutrient Imbalance and Tree Decline," papers presented at the NATO Workshop, Toronto, Canada, May 12-17, 1985.

61

51. Prinz et al., "Responses of German Forests in Recent Years"; Nico van Breeman, "Acidification and Decline of Central European Forests," *Nature*, May 2, 1985; for a technical overview of damage mechanisms see T.T. Kozlowski and Helen A. Constantinidou, "Responses of Woody Plants to Environmental Pollution," *Forestry Abstracts*, January 1986.

52. Tomas Paces, "Sources of Acidification in Central Europe Estimated from Elemental Budgets in Small Basins," *Nature*, May 2, 1985.

53. C.S. Holling, "Resilience of Ecosystems: Local Surprise and Global Change," in Malone and Roederer, eds., *Global Change*; Paces, "Sources of Acidification in Central Europe"; see also F.H. Bormann, "Air Pollution and Forests: An Ecosystem Perspective," *BioScience*, July/August 1985.

54. Andrew Csepel, "Czechs and the Ecological Balance," *New Scientist*, September 27, 1984; Deumling, private communication, March 1986; Von H. Steinlin, "Holzproduzierende Forstwirtschaft," presented at Conference on Forestry Management: Supplier of Raw Materials and the Environmental Factor, Göttingen, West Germany, November 14-15, 1983; Von H. Steinlin, "Waldsterben und Raumordnung," prepublication manuscript, Albert-Ludwigs University, Freiburg, West Germany, 1986.

55. Ewers et al., "Zur Monetarisierung der Waldschäden in der Bundesrepublik Deutschland."

56. Peter B. Reich and Robert G. Amundson, "Ambient Levels of Ozone Reduce Net Photosynthesis in Tree and Crop Species," *Science*, November 1, 1985.

57. See Sandra Postel, "Protecting Forests," in Lester R. Brown et al., *State of the World 1984* (New York: W. W. Norton & Co., 1984); Richard J. Meislin, "Mexico City's Flora Finds Life Too Foul to Bear," *The New York Times*, September 12, 1983.

58. Federal Ministry of Food, Agriculture and Forestry, "1985 Forest Damage Survey," Bonn, West Germany, October 1985; Deumling, private communication, October 1985.

59. Michael Castleman, "Toxics and Male Infertility," *Sierra,* March/April 1985.

60. For historical sketches of air pollution control see Erik P. Eckholm, *Down to Earth* (New York: W.W. Norton & Co., 1982) and Edwin S. Mills, *The Economics of Environmental Quality* (New York: W.W. Norton & Co., 1978).

61. OTA, *Acid Rain and Transported Air Pollutants.*

62. The Organisation for Economic Co-operation and Development (OECD), *State of the Environment* (Paris: 1985); NRC, *Epidemiology and Air Pollution* (Washington, D.C.: National Academy Press, 1985).

63. Brazil figures from NRC, *Epidemiology and Air Pollution; The State of India's Environment 1982* (New Delhi: Centre for Science and the Environment, 1982); Dianwu Zhao and Bozen Sun, "Air Pollution and Acid Rain in China," *Ambio,* Vol. 15, No. 1, 1986; Shanghai reference from Michael Weisskopf, "Shanghai's Curse: Too Many Fight for Too Little," *The Washington Post,* January 6, 1985.

64. Vander, *Nutrition, Stress, and Toxic Chemicals;* OECD, *State of the Environment.*

65. Joel Schwartz et al., *Costs and Benefits of Reducing Lead in Gasoline: Final Regulatory Impact Analysis* (Washington, D.C.: U.S. Government Printing Office, 1985); metal uptake by children from Vander, *Nutrition, Stress, and Toxic Chemicals;* figure of 675,000 from NRC, *Epidemiology and Air Pollution;* see also Richard Rabin, "Lead Poisoning: Silent Epidemic," *Science for the People,* July/August 1985.

66. Chanel Ishizaki and Juan Urich, "Mercury Contamination of Food: A Venezuelan Case Study," *Interciencia,* July/August 1985; Tri-Academy Committee on Acid Deposition, *Acid Deposition: Effects on Geochemical Cycling and Biological Availability of Trace Elements* (Washington, D.C.: National Academy Press, 1985); Don L. Johnson and Quincy Dadisman, "Acid Helps Mercury Contaminate Lakes," *Milwaukee Sentinel,* October 2, 1985; Eugeniusz Pudlis, "Poland: Heavy Metals Pose Serious Health Problems," *Ambio,* Vol. 11, 1982.

67. Tri-Academy Committee, *Acid Deposition: Effects on Geochemical Cycling;* Thomas H. Maugh II, "Acid Rain's Effects on People Assessed," *Science,* December 21, 1984.

68. Swedish Ministry of Agriculture, *Acidification Today and Tomorrow* (Stockholm: 1982); OTA, *Acid Rain and Transported Air Pollutants.*

69. Magda Havas et al., "Red Herrings in Acid Rain Research," *Environmental Science & Technology*, Vol. 18, No. 6, 1984; Bernhard Ulrich, "Dangers for the Forest Ecosystem Due to Acid Precipitation," translated for the EPA by Literature Research Company, Annandale, Va., undated; Maugh, "Acid Rain's Effects on People Assessed."

70. Daniel P. Perl, "Relationship of Aluminum to Alzheimer's Disease," *Environmental Health Perspectives*, Vol. 63, 1985; Daniel P. Perl et al., "Intra-neuronal Aluminum Accumulation in Amyotrophic Lateral Sclerosis and Parkinsonism-Dementia of Guam," *Science*, September 10, 1982.

71. Perl, "Relationship of Aluminum to Alzheimer's Disease."

72. James O. Schreck, *Organic Chemistry: Concepts and Applications* (Saint Louis, Mo.: The C.V. Mosby Company, 1975); "The Quest for Chemical Safety," *International Register of Potentially Toxic Chemicals Bulletin*, May 1985; Michael Shodell, "Risky Business," *Science '85*, October 1985.

73. Philip M. Boffey, "After Years of Cancer Alarms, Progress Amid the Mistakes," *The New York Times*, March 20, 1984; William U. Chandler, *Banishing Tobacco* (Washington, D.C.: Worldwatch Institute, 1986); Arthur J. Vander, The University of Michigan Medical School, Ann Arbor, private communication, April 1986.

74. NRC, *Toxicity Testing* (Washington, D.C.: National Academy Press, 1984); see also "Troubling Times with Toxics," *National Wildlife*, January 1986.

75. NRC, *Toxicity Testing.*

76. Charles Benbrook, Executive Director, Board on Agriculture, National Academy of Sciences, private communication, Washington, D.C., May 1986; see Philip Shabecoff, "Pesticide Control Finally Tops the E.P.A.'s List of Most Pressing Problems," *The New York Times*, March 6, 1986.

77. Ian C.T. Nisbet and Nathan J. Karch, *Chemical Hazards to Human Reproduction* (Park Ridge, N.J.: Noyes Data Corporation, 1983).

78. Janet Raloff, "Dioxin: Is Everyone Contaminated?" *Science News*, July 13, 1985.

79. Raloff, "Dioxin: Is Everyone Contaminated?"; Thomas H. Umbreit et al., "Bioavailability of Dioxin in Soil from a 2,4,5-T Manufacturing Site," *Science,* April 25, 1986; Susan Okie, "Dioxin May Weaken Ability to Fight Disease," *The Washington Post,* April 18, 1986; Janet Raloff, "Infant Dioxin Exposures Reported High," *Science News,* April 26, 1986.

80. Ralph C. Dougherty et al., "Sperm Density and Toxic Substances: A Potential Key to Environmental Health Hazards," reprint from J.D. McKinney, *Environmental Health Chemistry—The Chemistry of Environmental Agents as Potential Human Hazards* (Ann Arbor: Ann Arbor Science Publishers, Inc., 1980).

81. Dougherty et al., "Sperm Density and Toxic Substances"; Thomas H. Maugh II, "Tracking Exposure to Toxic Substances," *Science,* December 7, 1984; Castleman, "Toxics and Male Infertility."

82. Vander, *Nutrition, Stress, and Toxic Chemicals.*

83. Ibid.

84. Roger Lewin, "Parkinson's Disease: An Environmental Cause?" *Science,* July 19, 1985; for a fascinating sketch of how the link was discovered, see J. William Langston, "The Case of the Tainted Heroin," *The Sciences,* January/February 1985.

85. Emissions rates based on data from Ralph Rotty, Institute for Energy Analysis, Oak Ridge Associated Universities, Oak Ridge, Tenn., made available May 1986.

86. Villach conference statement, "An Assessment of the Role of Greenhouse Gases in Climate Variations"; Eike Röhling and Jochen Mohnfeld, "Energy Policy and the Energy Economy in FR Germany," *Energy Policy,* December 1985.

87. Peter Bunyard, "Sweden—Choosing the Right Energy Path," *The Ecologist,* Vol. 16, No. 1, 1986; Röhling and Mohnfeld, "Energy Policy and the Energy Economy in FR Germany."

88. "California Adopts Efficiency Standards," *The Energy Daily,* December 20, 1984; Vic Reinemer, "Progress—After 10 Years—on Appliance Efficiency Standards," *Public Power,* September/October 1985; Patricia Rollin and Jan Beyea, "US Appliance Efficiency Standards," *Energy Policy,* October 1985.

89. William U. Chandler, *Materials Recycling: The Virtue of Necessity* (Washington, D.C.: Worldwatch Institute, 1983); Richard Porter and Tim Roberts, eds., *Energy Savings by Wastes Recycling* (New York: Elsevier Science Publishing Co., Inc., 1985).

90. John A. Laurmann, "The Role of Gas in the Emission of Carbon Dioxide and Future Global Climate Change," *Gas Research Insights*, July 1985; American Gas Association, "A Review of the 'Joint Report of the Special Envoys on Acid Rain,'" Arlington, Va., January 1986.

91. Röhling and Mohnfeld, "Energy Policy in FR Germany"; see Brown et al., *State of the World* for the years indicated for discussions of nuclear power (1984 and 1986) and renewable energy (1984 and 1985); U.S. power projects from Christopher Flavin, "Reforming the Electric Power Industry," in Brown et al., *State of the World 1986*; Thomas Land, "Sweden to Go Non-Nuclear," *Worldpaper*, January 1986.

92. Commitment to sulfur reductions from "ECE Countries to Reduce Sulfur Emissions," *UNEP News*, September/October 1985.

93. Sand, "The Vienna Convention is Adopted"; EPA, "Analysis of Strategies for Protecting the Ozone Layer"; status of negotiations from EPA, "Stratospheric Ozone Protection Plan," *Federal Register*, January 10, 1986.

94. World Resources Institute, "The World's Tropical Forests: A Call for Accelerated Action," draft, Washington, D.C., June 1985.

95. Trefry et al., "A Decline in Lead Transport by the Mississippi River"; Schwartz et al., "Costs and Benefits of Reducing Lead in Gasoline."

96. Schwartz et al., "Costs and Benefits of Reducing Lead in Gasoline"; Warren Brown, "End Nears for Leaded Gasoline—And Bargain Fuel Prices," *The Washington Post*, December 29, 1986; Robert McDonald, "European Ministers Set Timetable for Auto Emission Standards," *World Environment Report*, April 17, 1985.

97. NRC, *Toxicity Testing*; David W. Schnare, "Examining the Toxicology of Low-Level Exposures: The Approaches and the Literature," presented at annual meeting of the American Association for the Advancement of Science, Los Angeles, Calif., May 1985.

65

66

98. NRC, *Toxicity Testing;* J. Clarence Davies, "Coping with Toxic Sub stances," *Issues in Science and Technology,* Winter 1985.

99. Michael Dover, "Getting Off the Pesticide Treadmill," *Technology Review,* November/December 1985; William F. Allman, "Pesticides: An Unhealthy Dependence?" *Science '85,* October 1985; see also Michael J. Dover and Briar A. Croft, "Pesticide Resistance and Public Policy," *BioScience,* February 1986

100. R.L. Frisbie and P.L. Adkisson, "IPM: Definitions and Current Status ir U.S. Agriculture," in Marjorie A. Hoy and Donald C. Herzog, *Biologica Control in Agricultural IPM Systems* (Orlando, Fla.: Academic Press, Inc. 1985).

101. Insufficient support cited in Dover, "Getting Off the Pesticide Tread mill."

102. Bruce Piasecki, ed., *Beyond Dumping: New Strategies for Controlling Toxi Contamination* (Westport, Conn.: Quorum Books, 1984).

103. OTA, *Technologies and Management Strategies for Hazardous Waste Contro* (Washington, D.C.: U.S. Government Printing Office, 1983); Joel Hirschhorn "Emerging Options in Waste Reduction and Treatment: A Market Incentive Approach," in Piasecki, ed., *Beyond Dumping;* Ron Wyden, "A Tax to Figh Toxic Waste," *The New York Times,* January 17, 1986.

104. NRC, *Global Change in the Geosphere-Biosphere;* Villach conference state ment, "An Assessment of the Role of Greenhouse Gases in Climate Vari ations."

SANDRA POSTEL is a Senior Researcher with Worldwatch Institute She is author of Worldwatch papers on water conservation and ai pollution effects on forests, and coauthor of *State of the World 1986* She studied geology and political science at Wittenberg University and resource economics and policy at Duke University.

THE WORLDWATCH PAPER SERIES

No. of
Copies

39. **Microelectronics at Work: Productivity and Job** Economy by Colin Norman.
40. **Energy and Architecture: The Solar and Conse** by Christopher Flavin.
41. **Men and Family Planning** by Bruce Stokes.
42. **Wood: An Ancient Fuel with a New Future** by Nigel Smith.
43. **Refugees: The New International Politics of Displacement**

Deudney.

Newland.
d Pamela

ewland.
ey.
wer

. Brown.

Deudney
Chandler

ra Postel
Chandler
r Brown and

cale

by

d

.
.
tel
Pollock
dra

DATE DUE

OCT 3 0 198

Bulk Co

Calenda

Make ch
1776 Ma

Washington, D.C. 20036 USA

Enclosed is my check for U.S. $ _____

name _____

address _____

city _____ state _____ zip/country